# GREATDESIGNS

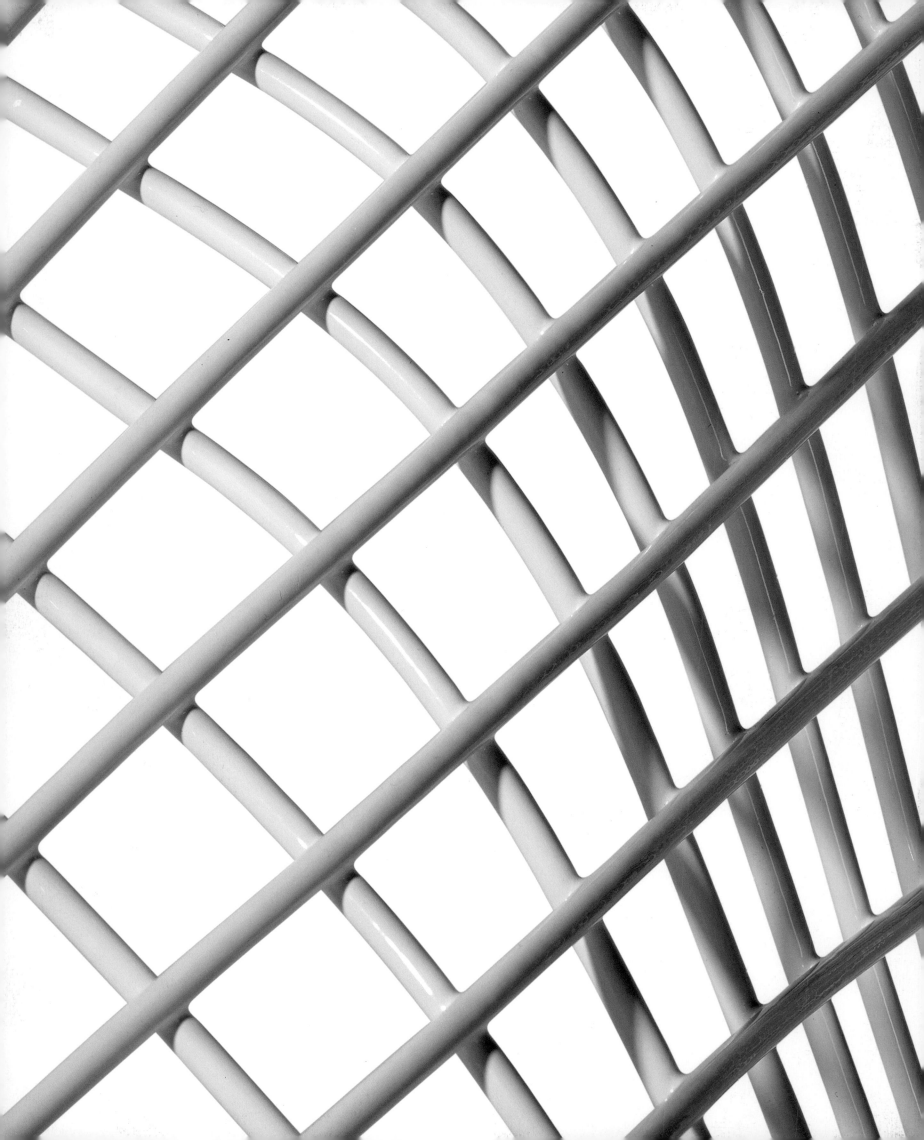

# GREATDESIGNS

# CONTENTS

Looking at design     6

## 1920–1929

| | |
|---|---|
| **Bentwood chair Model 14** | 14 |
| Michael Thonet | |
| **Fruit wallpaper** | 16 |
| William Morris | |
| **Silver teapot** | 20 |
| Christopher Dresser | |
| **Swiss Army Knife** | 22 |
| Karl Elsener | |
| **Moët et Chandon poster** | 24 |
| Alphonse Mucha | |
| **Tiffany lamp** | 26 |
| Clara Driscoll, | |
| Louis Comfort Tiffany | |
| **Hill House chair** | 28 |
| Charles Rennie Mackintosh | |
| **Flatware** | 30 |
| Josef Hoffmann, | |
| Wiener Werkstätte | |
| **Electric kettle** | 32 |
| Peter Behrens | |
| **Red Blue chair** | 34 |
| Gerrit Rietveld | |

| | |
|---|---|
| **L'Élégance perfume bottle** | 38 |
| René Lalique | |
| **Silver pitcher** | 40 |
| Johan Rohde | |
| **AGA cooker** | 42 |
| Gustaf Dalén | |
| **Dobrolet poster** | 44 |
| Alexandr Rodchenko, | |
| Varvara Stepanova | |
| **Bauhaus poster** | 46 |
| Joost Schmidt | |
| **E1027 occasional table** | 50 |
| Eileen Gray | |
| **Cité armchair** | 52 |
| Jean Prouvé | |
| **Nord Express poster** | 54 |
| A.M. Cassandre | |
| **LC4 chaise longue** | 56 |
| Le Corbusier, Charlotte Perriand, | |
| Pierre Jeanneret | |
| **Barcelona chair** | 58 |
| Ludwig Mies van der Rohe | |

## 1930–1939

| | |
|---|---|
| **Moka Express coffee-maker** | 62 |
| Alfonso Bialetti | |
| **London Underground map** | 64 |
| Harry Beck | |
| **Ericsson telephone DHB 1001** | 68 |
| Jean Heiberg, Christian Bjerknes | |
| **Pencil sharpener** | 70 |
| Raymond Loewy | |
| **Anglepoise lamp** | 72 |
| George Carwardine | |
| **Ekco radio AD65** | 74 |
| Wells Coates | |
| **Savoy vase** | 76 |
| Alvar Aalto | |
| **Kodak Bantam Special** | 78 |
| Walter Dorwin Teague | |
| **Volkswagen Beetle Model 1300** | 80 |
| Ferdinand Porsche | |
| **Knuten candelabra** | 84 |
| Josef Frank | |
| **B.K.F. chair** | 86 |
| Antonio Bonet, Juan Kurchan, | |
| Jorge Ferrari Hardoy | |

## 1940–1954

| | |
|---|---|
| **Emerson Patriot radio** | 90 |
| Norman Bel Geddes | |
| **Harper's Bazaar magazine** | 92 |
| Alexey Brodovitch | |
| **Tupperware** | 96 |
| Earl Tupper | |
| **Vespa** | 98 |
| Corradino D'Ascanio | |
| **Penguin paperback covers** | 102 |
| Jan Tschichold, Edward Young | |
| **Coffee table** | 104 |
| Isamu Noguchi | |
| **Fazzoletto vase** | 106 |
| Paolo Venini, Fulvio Bianconi | |
| **Atomic wall clock** | 108 |
| George Nelson | |
| **Calyx furnishing fabric** | 110 |
| Lucienne Day | |
| **Festival of Britain symbol** | 114 |
| Abram Games | |
| **Birch platter** | 116 |
| Tapio Wirkkala | |
| **Kilta tableware** | 118 |
| Kaj Franck | |
| **Diamond armchair** | 120 |
| Harry Bertoia | |
| **Pride cutlery** | 122 |
| David Mellor | |
| **Fender Stratocaster** | 124 |
| Leo Fender | |
| **M3 Rangefinder camera** | 128 |
| Leica Camera AG | |
| **Butterfly stool** | 132 |
| Sori Yanagi | |

LONDON, NEW YORK, MUNICH,
MELBOURNE, DELHI

WRITTEN BY

**Philip Wilkinson**

| | | | |
|---|---|---|---|
| Senior Editor | Angela Wilkes | Jacket Design Manager | Sophia M.T.T. |
| Senior Art Editor | Gillian Andrews | Managing Editor | Stephanie Farrow |
| Editors | Anna Kruger | Senior Managing Art Editor | Lee Griffiths |
| | Hugo Wilkinson | Publisher | Andrew Macintyre |
| Designer | Phil Gamble | Art Director | Phil Ormerod |
| Producer, Pre-production | Lucy Sims | Associate Publishing Director | Liz Wheeler |
| Production Controller | Mary Slater | Publishing Director | Jonathan Metcalf |
| Picture Research | Myriam Mégharbi, | | |
| | Sarah Smithies | | |
| Jacket Designer | Jess Bentall | | |
| Jacket Design Editor | Manisha Majithia | | |

## 1955–1959

**The Man with the Golden Arm poster**   136
Saul Bass

**Citroën DS**   138
Flaminio Bertoni

**Apple vase**   142
Ingeborg Lundin

**Phonosuper SK4**   144
Dieter Rams, Hans Gugelot

**Mirella sewing machine**   148
Marcello Nizzoli

**Tulip table**   150
Eero Saarinen

**Lounge Chair 670 and ottoman**   152
Charles and Ray Eames

**Wall clock**   156
Max Bill

**Helvetica typeface**   158
Max Miedinger

**Egg chair**   162
Arne Jacobsen

**PH Artichoke lamp**   164
Poul Henningsen

**Mono-a flatware**   166
Peter Raacke

**Austin Seven Mini**   168
Alec Issigonis

**Cadillac series 62**   172
Harley Earl

**Panton chair**   176
Verner Panton

## 1960–1979

**Arco lamp**   180
Achille and Pier Giacomo Castiglione

**Moulton bicycle**   182
Alex Moulton

**Unikko fabric**   186
Maija Isola

**Grillo folding telephone**   188
Marco Zanuso, Richard Sapper

**Kodak Instamatic 33**   190
Kenneth Grange

**Valentine typewriter**   192
Ettore Sottsass, Perry A. King

**Revolving cabinet**   196
Shiro Kuramata

**Beogram 4000**   198
Jacob Jensen

**Tizio desk lamp**   200
Richard Sapper

**Wiggle chair**   202
Frank Gehry

**Munich Olympic Games pictograms**   204
Otl Aicher

**Suomi tableware**   206
Timo Sarpaneva

## 1980–1995

**Carlton bookcase**   210
Ettore Sottsass, Memphis Group

**The Face magazine**   212
Neville Brody

**Whistling Bird kettle**   214
Michael Graves

**Wood chair**   216
Marc Newson

**Carna wheelchair**   218
Kazuo Kawasaki

**Bookworm bookshelf**   220
Ron Arad

**85 Lamps chandelier**   222
Rody Graumans, Droog Design

**Vermelha chair**   224
Fernando and Humberto Campana

**Dyson DC01 vacuum cleaner**   226
James Dyson

**Verdana typeface**   228
Matthew Carter

## 1996 TO PRESENT

**Brick shelf system**   232
Ronan and Erwan Bouroullec

**Fjord Relax armchair**   234
Patricia Urquiola

**Garland light**   236
Tord Boontje

**Lover sofa**   238
Pascal Mourgue

**Miura stackable stool**   240
Konstantin Grcic

**Evolute table lamp**   242
Matali Crasset

**Spun chair**   244
Thomas Heatherwick

**Apple iPad**   246
Jonathan Ive, Apple Industrial Design

**Masters chair**   248
Philippe Starck

**Index**   250

**Acknowledgments**   255

First published in Great Britain
in 2013 by Dorling Kindersley Limited
80 Strand,
London WC2R 0RL

Penguin Group (UK)
2 4 6 8 10 9 7 5 3 1
001 – 184768 – Sept/2013

Copyright © 2013 Dorling Kindersley Limited

Printed and bound in China by
Leo Paper Products Ltd

Discover more at
**www.dk.com**

A CIP catalogue record for this book
is available from the British Library
ISBN 978-1-4093-1941-2

# Looking at design

**Design is part of our everyday lives.** Nearly every object around us at home and work has been designed. Whether it is a table or a camera, a bicycle or a computer, someone has planned exactly what it should look like, how it should work, and what materials it should be made of, creating drawings, plans, and instructions for its construction. This is how nearly all manufactured objects are created today. Designers shape our environment, down to the smallest details, and it makes them some of the most influential people in the world.

The profession of designer slowly evolved in the 18th and 19th centuries with the rise of industry. Before then, most of the things that people used were handmade by craftsmen. During the Industrial Revolution, more and more things were mass-produced in factories or mills, and designers were employed to draw or model products that would then be moulded, cut out, assembled, and finished by machines. The middle classes were becoming increasingly prosperous, and the growing demand for furniture, lamps, and tableware to furnish their homes and work environments led to a successful partnership between designers and manufacturers from the late 19th century onwards.

Some designers were initially hostile towards mass production, but many of them welcomed the opportunities it gave them to create items that were both useful and beautiful using the latest materials and new, faster production techniques. Fruitful collaborations between design and manufacturing led to the innovative electrical appliances designed by Peter Behrens for AEG and Dieter Rams for Braun; to chairs as diverse as Mies van der Rohe's Barcelona chair and Harry Bertoia's Diamond chair; and to great cars, from the Citroën DS to the Mini.

Looking at design is a personal experience, but one that is enriched by broader knowledge – the more you know about it, the closer you look and the more you notice and enjoy. This book takes you on a guided tour of some of the finest designs from the last 150 years, ranging from the classic 19th-century bentwood chair by Thonet to the latest creations of Thomas Heatherwick, Jonathan Ive, and Philippe Starck. It picks out the key details of each design, showing how form, materials, colour, and techniques have been used, and provides fascinating insights into the stories and characters behind it. Good, affordable designs are all around us – knowing more about why and how they have been created helps us to appreciate what makes some designs truly great.

**AEG kettle,** Peter Behrens, 1908

**Emerson Patriot radio,** Norman Bel Geddes, 1940

**Barcelona Chair,** Ludwig Mies van der Rohe, 1929

# Developing design

By the end of the 19th century, many designers had started working closely with manufacturers, creating items for mass production. Some of these early industrial designers were little-known figures who nevertheless made a huge impact, such as Karl Elsener, the designer of the 1897 Swiss Army Knife. A few were versatile professional designers, such as Peter Behrens, whose prolific and varied work for the German company AEG in the early 20th century earned him recognition as the first true industrial designer. Behrens was just as happy whether he was designing a kettle, a logo, or even an entire factory. The way in which he adapted to the needs of mass production paved the way for the designers who came after him.

## ART AND TECHNOLOGY

The challenge for early designers was to combine the ideas and skills of the artist with the materials and production techniques of industry. In the 1920s, the Russian Constructivists successfully combined art and technology, and produced powerful, modern-looking work ranging from graphics to textiles. This was a diverse movement enriched by the work of internationally known artist-designers such as Alexandr Rodchenko and El Lissitsky. The Modernist German Bauhaus school, founded in 1919, also taught artists to combine rigorous design with modern, industrial ideas. It espoused a functional, minimalist approach and embraced the doctrine of "truth to materials", celebrating the qualitites of materials rather than concealing them beneath applied ornament or colour. The Bauhaus was hugely influential, especially after it was forced to close under the Nazi regime, when its teachers and designers relocated, taking its ideas all over the world.

Meanwhile, in France, Modernist designers such as Le Corbusier and Charlotte Perriand pursued the idea that form follows function, and designed furniture inspired by new, industrial materials such as tubular steel. The American designers Raymond Loewy and Norman Bel Geddes had a different outlook. They used industrial materials, including newly developed plastics, to shape their products or give them vibrant colours, embracing fashions such as streamlining. Because they had a large and expanding commercial market and access to mass production, these designers brought modernity to the public, helping to democratize design. Although functional, their designs placed a strong emphasis on appearance, and the streamlined cars, electrical goods, and office equipment of the 1930s and 1940s fired people's imaginations and seemed to promise a rosy future of bold new shapes, bright colours, and cutting-edge technology.

## POSTWAR DESIGN

Designers were aware of the need for a fresh start after World War II. Some, like the great German designer Dieter Rams, drew on the Modernist tradition. Minimalist design, a restricted colour palette, and carefully designed controls and switches are typical features of the work he produced for Braun. Other designers used soft edges and gentle curves to create elegant pieces, such as the Butterfly stool by Sori Yanagi, or Charles and Ray Eames's Lounge chair.

National identities were often reflected in approaches to design. Whereas German designers tended towards functionalism, Italians produced designs that were more organic in appearance and often vividly coloured. The UK experienced a renaissance in design, encouraged by the 1951 Festival of Britain, and the work of British designers such as Lucienne Day and Kenneth Grange won international acclaim. In Scandinavia, designers drew on a strong craft tradition and created designs characterized by purity of form, ranging from the glassware produced by the Finnish company Iittala to the curvaceous furniture of Arne Jacobsen.

Bright colours became even more fashionable in the 1960s, when Pop art had a decisive influence on design. Restrained functionalism gave way to experimentation and bold design, fuelled by popular culture, a consumer boom, and the increasing spending power of the young. The influence of Pop art spread to everything from fabrics and light fittings to moulded plastic furniture. Radical designs, such as the Austin Mini and the Moulton bicycle, both of which began as engineering projects, seemed to epitomize the spirit of the decade and were featured in many fashion shoots and contemporary films.

## AGAINST MODERNISM

Postmodernism, a dominant design movement of the late 20th century, also used bright colours in provocative ways. Designs such as the Carlton bookcase by Ettore Sottsass creatively combined bright colours and unusual shapes, overturning conventional ideas about what a piece of furniture should look like. Instead of being severely functional or gently organic, Postmodern design could be playful, provocative, and surprising. Unlike their Modernist predecessors, Postmodern designers deliberately borrowed motifs and ornamentation from the past –whether they were classical columns or jazzy patterns drawn from 1950s laminates – and gleefully incorporated them into their own designs. By deliberately pursuing unpredictable paths, Postmodern designers helped to liberate design, freeing it from the lingering influence of Modernism, which many people had come to regard as both dominant and restricting.

## DIGITAL DESIGN

Designers today use a wider array of materials, colours, and production methods than ever before. The manufacturing potential of plastics has led to innovative machines, such as the Dyson vacuum cleaner, radical new pieces of furniture, and lightweight, portable, stylish digital appliances. Computer technology has helped to transform design in many different ways, from computer-aided design to new production techniques that enable designers to send manufacturers the digital files for their products. Despite technological advances, however, the fundamentals of design remain essentially the same as they were 100 years ago. Designers analyse what is required by the client and user, search for creative solutions, and come up with attractive, functional designs that can be manufactured with the available technology for the right price. They approach each project with the same principles and creativity as they have always done and we can look forward with confidence to a future filled with ever more ingenious great designs.

# "It's very easy to be different, but very difficult to be better"

**JONATHAN IVE**

**Austin Seven Mini,** Alec Issigonis, 1959

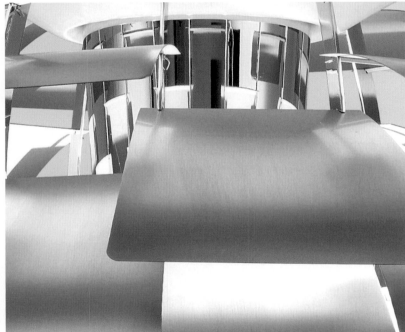

**Carlton bookcase,** Ettore Sottsass, 1981

**Vermelha chair,** Fernando and Humberto Campana, 1993

- **Bentwood chair Model 14** Michael Thonet
- **Fruit wallpaper** William Morris
- **Silver teapot** Christopher Dresser
- **Swiss Army Knife** Karl Elsener
- **Moët et Chandon poster** Alphonse Mucha
- **Tiffany lamp** Clara Driscoll, Louis Comfort Tiffany
- **Hill House chair** Charles Rennie Mackintosh
- **Flatware** Josef Hoffmann, Wiener Werkstätte
- **Electric kettle** Peter Behrens
- **Red Blue chair** Gerrit Rietveld

# Bentwood chair Model 14

1859 ▪ FURNITURE ▪ BEECHWOOD AND CANE ▪ GERMANY

SCALE

## MICHAEL THONET

**Prized for its combination of lightness** and gentle, organic curves, the Thonet Model 14 bentwood chair is one of the most successful pieces of furniture ever made. From the 1860s onwards, the chair was widely used in European bistros and cafés and, owing to its ease of assembly from ready-made components, it was soon exported worldwide.

The Model 14 chair was produced using a revolutionary process developed by its designer, Michael Thonet, who used the heat from steam to bend its solid beechwood frame into shape. Before this, cabinet-makers had to carve or cut wood in order to produce curves, but the steam-bending process enabled Thonet to produce the single, sinuous curve formed by the two rear legs and the chair back from just one piece of timber. Additional bentwood components – the two front legs, another loop to reinforce the back, and a circle to support the seat – completed the frame, and Thonet joined the pieces together using screws rather than glue and carving. The result was a simple, strong chair that was elegant, easy to assemble, and affordable. It was such a success that between 1859 and 1930 about 30 million chairs were produced. Admired by architects such as Le Corbusier, the Model 14 made the Thonet company internationally famous, and the chair spawned several variations from other manufacturers. The original model is still produced in a modified form today.

The back varies in thickness according to structural needs

The seat is positioned flush with the frame

Using cane keeps the chair light

The legs splay slightly to give the chair added stability

### MICHAEL **THONET**

**1796–1871**

German-born cabinet-maker Michael Thonet set up a furniture business in Boppard am Rhein in 1819, and began experimenting with bending wood in about 1830. His early furniture found favour in Vienna, Austria, where he produced work for the imperial court as well as developing mass-market products. In 1853, Thonet handed his business over to his five sons, and in 1859, they began manufacturing the Model 14 chair in their factory in Koryčany, Moravia. Still bearing the name of its founder, the company continues to manufacture beautifully designed furniture.

# Visual tour

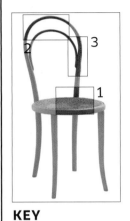

**KEY**

▶ **SEAT** Woven cane was often used for the seats of early Model 14 chairs. Its comfort and light weight were appreciated by café owners, as well as its practicality: spilt liquids flowed straight though the cane to the floor, where staff could mop them up.

1

▶ **CHAIR BACK**
The curved back and reinforcing loop provide just enough support for the user and contrast with the ornate backs of many 19th-century chairs. The shaped wood also makes the chair easy to pick up and carry.

2

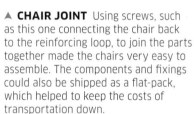

3

▲ **CHAIR JOINT** Using screws, such as this one connecting the chair back to the reinforcing loop, to join the parts together made the chairs very easy to assemble. The components and fixings could also be shipped as a flat-pack, which helped to keep the costs of transportation down.

## ON **DESIGN**

The Thonet company built up a large catalogue of bentwood, including chairs, tables, washstands, and cradles. There were many variations on all of these pieces, and bentwood components were used both structurally and ornamentally. Some of the most enduring designs were rocking chairs. Adding numerous bentwood braces and decorative flourishes made these chairs into elaborate studies in coordinated curves. One of the most complex of all was the reclining rocker or lounger (below), in which the extra length provides scope for long pieces of beechwood. These curve back and forth in spirals and arabesques that anticipate the whiplash curves of the Art Nouveau style.

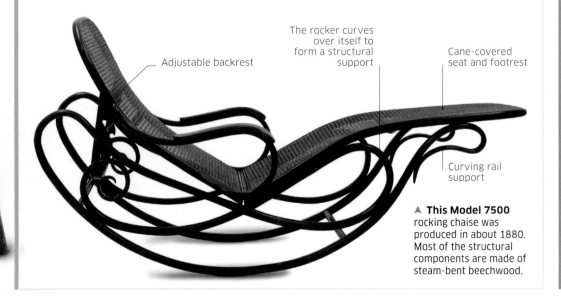

Adjustable backrest

The rocker curves over itself to form a structural support

Cane-covered seat and footrest

Curving rail support

▲ **This Model 7500** rocking chaise was produced in about 1880. Most of the structural components are made of steam-bent beechwood.

## ON **PRODUCTION**

To create bentwood chair frames, Thonet Brothers used a range of specially made iron moulds mounted on solid stands. Craftsmen forced ready-shaped, steam-heated lengths of beech into these moulds, where they were left to dry for around 20 hours at about 70°C (21°F). In the 1850s, the company was given an exclusive 10-year licence to bend wood in this way, so they expanded the business, opening several factories and exporting furniture around the world. By the time the licence expired in 1869, Thonet was well established and it went on to enjoy continuing success.

▲ **Thonet factory, late 1850s** Working in pairs, craftsmen bend heated wood around metal frames.

# Fruit wallpaper

1862-66 ■ WALL COVERING ■ BLOCK-PRINTED PAPER ■ UK

SCALE

## WILLIAM MORRIS

**British designer, writer, and campaigner William Morris** was famed for his furniture and stained-glass designs, as well as his writings on many subjects from conservation to socialism. Among his most influential work was a series of wallpapers he designed in the 1860s and 1870s. Fruit, one of the earliest, is an arrangement of pomegranate, orange, lime, and peach tree branches, all bearing leaves, fruit, and flowers. It was strikingly different from the Victorian wallpapers popular at the time because the pattern was treated in a flattened, stylized fashion, without the shading that made most contemporary wallpaper designs look three-dimensional, and set against a light background.

Like the British Pre-Raphaelite painters and the great writer and art critic John Ruskin, Morris drew inspiration from a pre-industrial era, particularly the Middle Ages,

when artists celebrated the beauty of the natural world and guilds of craftsmen made everything from a chair to a cathedral by hand. For his Fruit design, Morris drew flowers, fruit, and leaves from life, but he was also influenced by woodcuts of plants in 16th-century herbals, some of which he owned, as well as the depictions of flowers and fruit on medieval tapestries.

Morris's handmade wallpapers were not popular in Victorian times and were prohibitively expensive for most people. Over time, however, their popularity grew. By the 1890s, his wallpapers and textiles were seen in the homes of the cultured middle classes, and his designs enjoyed a revival in 1970s Britain. Fruit and other patterns are still made today, a lasting testament to the strength of Morris's design skills and evocation of the natural world.

"Have nothing in your house that you do not know to be useful, or believe to be beautiful"

**WILLIAM MORRIS**

▶ **Preliminary drawing**
Morris began by making watercolour drawings of his design to determine the balance of shapes and how the pattern repeats would work. This surviving design for Fruit (right) shows vigorous drawing, taken in part from Morris's tapestry designs, with olives (not included in the final paper) in the lower left quarter. Morris often collaborated with his associates and some experts have detected the hand of the architect Philip Webb, who worked closely with Morris, in this drawing.

### WILLIAM **MORRIS**

#### 1834-96

William Morris went to Oxford University to prepare for the priesthood, but he fell in love with medieval architecture and decided he would train as an architect. In 1861, he became one of the founders of Morris, Marshall, Faulkner & Co. (later known as Morris & Co.), a firm set up to design and manufacture fabrics, tapestries, furniture, stained glass, metalwork, and other decorative items. All were beautifully designed, made of the finest natural materials, and constructed using traditional techniques. Morris provided many of the designs himself, especially for the wallpapers and textiles. He was also a prolific poet and polemical writer, promoting socialism and the importance of traditional craftsmanship. Some of his books were hand-printed at his own publishing house, the Kelmscott Press.

# Visual tour

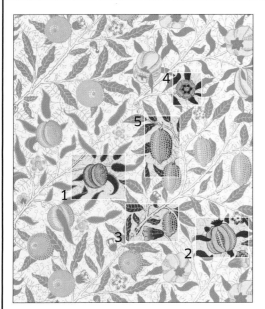

**KEY**

▼ **PEACH** The design uses a combination of pink, green, and light browns in a variety of tones to depict the subtle range of colour on the skin of the peach. The colour has been applied in small patches, giving an impression of the soft texture of the fruit's surface and depicting changes in tone. A strong, dark line indicates the pronounced "fold" in the fruit's rounded form.

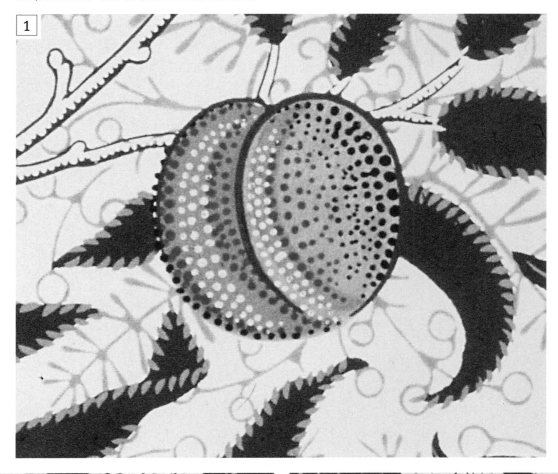

▼ **POMEGRANATE** Around a quarter of the pattern is made up of pomegranates, their greyish skins splitting to reveal rows of glossy red seeds. Morris would have known from his reading of medieval literature that the pomegranate is an ancient symbol of eternal life and its numerous seeds represent fertility. The fruit is an important element in the design, which is sometimes known as Pomegranate.

▲ **BRANCH AND LEAF** In contrast to the dark green leaves that stand out against the pale background, the branch is shown in outline, as if in a pen-and-ink drawing. The knobbly shape of the slender branch adds form and texture. The cut end of the branch is realistic and was probably drawn from nature.

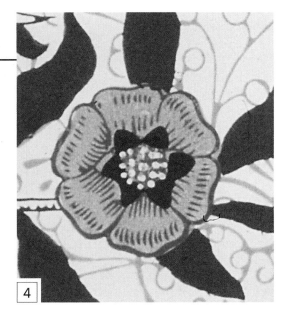

**4**

▲ **FLOWER** Morris loved medieval illuminated manuscripts, old woodcuts, and tapestries, and the form of this pomegranate flower is more likely to have been influenced by these sources rather than drawn from life. This flower is important in the design more for the way in which its petals provide a splash of rich, glowing colour than for its botanical accuracy.

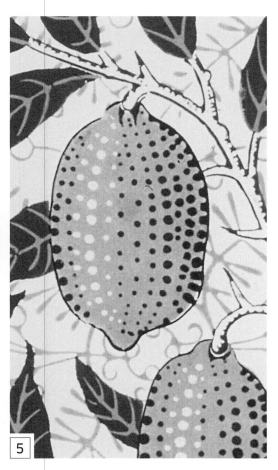

**5**

▲ **LIME** The group of limes is a good example of how realism and stylization blend in the wallpaper. The shape of the fruit, with its bent stem and the characteristic protrusion at the base, is drawn in a realistic manner. The rough texture of the skin, on the other hand, s represented by a mosaic of small squares of dark and light greens.

## IN **CONTEXT**

Morris's firm produced a wide range of wallpapers based on botanical subjects, mostly leaves and flowers. Some he designed himself, others were created by associates such as Kate Faulkner and Morris's daughter, May. Although all the designs are rich in detail, they are graphic and relatively simple compared with most Victorian wallpapers of the time. Morris succeeded in concealing the pattern repeat with skilful use of ornamentation.

He also departed from the conventions of Victorian interiors by restricting himself to one pattern. Many decorators of the period liked to hang several patterns on one wall – one for the dado, one for the main wall, another for the frieze, and yet another for the border. Morris preferred to use a single design from the skirting to the ceiling. The wallpapers in his own homes, including Kelmscott Manor (below), were hung in this way.

▲ **Morris wallpapers and tapestries at Kelmscott Manor**

## ON **DESIGN**

Morris's ideas on design were echoed by several other late 19th-century British craftsmen and designers. Influenced by the ideas of critic John Ruskin, who linked the quality of art and design with the social conditions that produced it, they turned their back on mechanization, believing that work crafted by hand would help produce not just better goods but a better society. This group, known as the Arts and Crafts Movement, included some of the

greatest designers of the period, including the architect C.F.A. Voysey, the metalwork designer C.R. Ashbee, and the ornamental designer Lewis F. Day. They embraced craftsmanship, the use of fine, natural materials, and ornamentation inspired by nature. Many of them worked together, forming guilds such as Ashbee's Guild of Handicraft that produced high-quality, handcrafted work in a wide range of fields, from furniture to stained glass.

◄ **Oak sideboard** Cabinet-maker Ernest Gimson moved to Gloucestershire in 1893 and, in collaboration with Sidney Barnsley, produced beautifully crafted, functional furniture that was architectural in form. Their work combined traditional techniques, respect for fine materials, and clean, elegant lines.

# Silver teapot

1880 ■ METALWARE ■ ELECTROPLATED NICKEL SILVER  AND EBONY ■ UK

## CHRISTOPHER DRESSER

SCALE

**In 1879, designer Christopher Dresser** began a series of projects for silver-plated tea sets. These designs, created for James Dixon and Sons of Sheffield, were well ahead of their time. Most 19th-century tableware was based on traditional, curved shapes, to which designers applied various forms of ornamentation derived from historical sources. Dresser's tea sets, by contrast, are exercises in pure, undecorated form. The most striking piece of all is this teapot, which has a square body set at an angle, like a diamond, and pierced by a square aperture in the centre. The legs and handle are perfectly straight and meet the body at 90° and 45° angles, so that the teapot looks like a study in linear geometry, and resembles a Modernist piece from the 1930s rather than something from the 1880s.

Dresser's aesthetic was partly influenced by the elegant simplicity of Japanese design, but this fascinating teapot owes its unusual form to his enthusiasm for strong, abstract shapes and patterns. He also applied principles derived from his study of nature, where he saw simple form and clear function as intrinsic to beauty. His innovative designs have a more pragmatic side too. Some of the elements – the handles and their brackets, for example – were used on more than one design

of teapot, thereby reducing production costs. Dresser was moving towards a more industrialized approach to design and manufacturing – indeed, he is sometimes seen as Europe's first industrial designer. His extraordinary designs were not, however, mass-produced and they remained the preserve of discerning middle-class customers, who appreciated them for their unusual shapes, polished surfaces, and quirky, distinctive appearance.

The lid is an integral part of the square form of the teapot

The central opening is a radical touch

### CHRISTOPHER **DRESSER**

#### 1834-1904

Christopher Dresser was born in Glasgow and attended the Government School of Design in London. He became a specialist in botany and insisted that designers should abstract and adapt botanical forms rather than copying them directly from nature. A successful designer of wallpapers, textiles, ceramics, and metalwork, he developed an interest in Japanese art and travelled to Japan, after which his designs became more minimalist and geometric. Dresser worked for many manufacturers and had a huge influence on British design that continued long after his death.

## ON **DESIGN**

Dresser's designs of the 1880s are exercises in paring an object down to its constituent parts. A toast rack becomes a series of straight lines; the body of a teapot can be a square, a cylinder, or a sphere; other items are made up of cones. This geometry goes hand in hand with functionality: all the component parts are clearly visible and the user can see instantly, for example, where the handle of an object ends and the body begins. There is no attempt to conceal the joint. Dresser even saw practical items such as a watering can (right) in terms of geometry. Its body and spout form two truncated cones connected by a semicircular handle.

▲ **Enamelled metal watering can**

# Visual tour

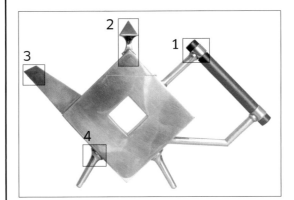

**KEY**

The ebony is practical as well as aesthetic, staying relatively cool

This angular support mirrors the line of the teapot's body

## "Maximum effect with minimum means"

**Christopher Dresser**

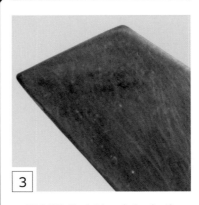

▲ **HANDLE AND SUPPORT** The ebony rod, set parallel to the body of the teapot, is held firmly in place by two metal tubes. These joints are good examples of the functionality that distinguished Dresser's designs.

◄ **FINIAL** The size and shape of the finial make it easy to remove the lid. Whereas most Victorian teapots had rounded finials, Dresser opted for a pyramid. Looked at in profile, its straight lines form a triangle that echoes the triangular shape of the lid.

▲ **SPOUT** Straight and simple, the angle of the spout counterbalances that of the handle on the other side of the teapot. The plain, rectangular opening does not have a pronounced lip and was designed to pour cleanly. Its polished silver finish matches that of the rest of the teapot.

▲ **FOOT AND JOINT** The splayed feet are round in profile and bulge slightly where they meet the body, breaking up the severe geometry of the teapot. Their simplicity forms a marked contrast to the elaborate scrolls and mouldings that were the hallmarks of Victorian design.

# Swiss Army Knife

1897 ▪ PRODUCT DESIGN ▪ STEEL AND HARDWOOD ▪ SWITZERLAND

SCALE

## KARL ELSENER

**Packing many functions into its compact red case**, the Swiss Army Knife is the archetypal pocket knife for millions of people. Yet when Karl Elsener began to produce the knives in the 1890s, he had a very specific market in mind, and did not envisage exporting them all over the world. In 1891, Elsener's company took over production of knives for the Swiss Army. At the time, Swiss soldiers were using a knife that contained the required tools – a blade, a screwdriver for taking apart the standard-issue army rifle, and a can opener – within a dark wooden casing. Elsener set about improving the design, and by using a special spring he found a way of installing tools and blades on both sides of the handle, so it became possible to include more functions. His Officer's Knife had two cutting blades, a screwdriver, a reamer (for punching holes), a tin-opener, and a corkscrew. The design fitted the brief perfectly, and was also adaptable. In the following decades, Elsener's company produced similar knives with different blades and tools to suit a variety of users, from travellers to hunters – the first toolkits that could be carried in a pocket.

Screwdriver for slot-headed screws

Large knife blade

Red is the standard colour, reflecting its use on the Swiss flag

Reamer

Small blade for finer cutting

The reamer folds into this space

Grip for pulling out the tin-opener

Suspension loop

### KARL **ELSENER**

#### 1860–1918

Born in Switzerland, Karl Elsener trained as a knife-maker, and specialized in making razors and surgical instruments. In 1884, he set up the Swiss Cutlery Guild to provide jobs in an area south of Geneva with high unemployment. In 1891, the Guild (later named Victorinox) was one of two firms contracted to make knives for the Swiss Army. Since Elsener's death the company has remained in the hands of his descendants.

# Visual tour

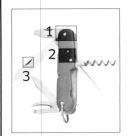

**KEY**

**► CURVED END** The knife has a semicircular end and all the edges are curved. These details make it feel comfortable in the hand, and means that there are no hard or sharp edges to catch on anything in the pocket.

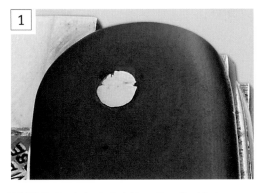

**1**

**2**

**◄ RIVETS** Steel rivets hold together the blades, spring, and body casing, as well as providing pivots for the tools. On the original knife, the rivets pass through the case and are visible on the outside – a piece of purely functional design.

**▼ INDENTATION FOR THUMBNAIL** Most of the tools, including the two cutting blades and screwdriver, have a small indentation, deeper in the centre and gently curved on one edge. It is easy to pull the tools out into the working position using a thumbnail.

Corkscrew

**3**

By the early 20th century, Elsener's Officer's Knives bore the distinctive shield and cross motif, based on the Swiss flag. New products included the *"Cure-pied"*, with its tool for cleaning horses' hooves, and an economical student's knife. Elsener renamed the company Victoria in 1909, after his mother, who had died that year.

In 1921, there was a further name change to Victorinox, incorporating the French term for stainless steel, *acier inoxydable*, or *inox* for short, which was then being used to make the blades.

**▲ Early catalogue pages in German and French**

## ON **DESIGN**

There is now a large range of Swiss Army Knives with stainless steel blades intended for diverse users. Models such as the Spartan, which offers a similar array of tools to those of the original Officer's Knife, are still made. For those who need more tools, there are models such as the Huntsman, which features a saw and scissors. Top-of-the-range products include the Swiss Champ, which has a vast assortment of blades, files, saws, pliers, and even a magnifying glass. Most models feature a toothpick and a pair of tweezers that slide into holes at one end of the case. And by means of an ingenious design, different functions are combined so that a bottle opener, screwdriver and wire stripper appear on a single "blade". Another company, Zenger (merged with Victorinox in 2005), produced a further range of Swiss Army Knives.

**▲ Victorinox Spartan knife**

**▲ Victorinox Huntsman knife**

**▲ Victorinox Swiss Champ knife**

# Moët et Chandon poster

1899 ▪ GRAPHICS ▪ COLOUR LITHOGRAPH ▪ FRANCE

SCALE

## ALPHONSE MUCHA

**Mucha's posters** are some of the most celebrated examples of Art Nouveau, the forward-looking art and design movement that swept across Europe during the 1890s. In his work for companies such as the Moët et Chandon champagne house, Mucha rejected the revivalist styles of the 19th century, turning instead to organically inspired decoration, flowing lines, and elaborate, ornamental typography. In Mucha's images, beautiful young women, often dressed in flowing gowns, are surrounded by flowers and foliage, and framed by sinuous curves.

This poster design for champagne is particularly striking. The elongated form of the woman, outlined in black, is set against an arched "window" with long panels – a sophisticated framing device that displays the name of the company and its product. The flattened background is a riot of colourful, stylized foliage and flowers that throws the figure into relief. Mucha's posters, including this one, were often printed on long sheets to emphasize the elegance of their female subjects and catch the eye. They proved very popular and Mucha had contracts with several major clients. He also designed posters for the actress Sarah Bernhardt and the dancer Loïe Fuller. Displayed in many of the major cities of Europe and widely reproduced in periodicals, these graphics ensured Mucha's reputation spread to the United States and helped to popularize the new style. Mucha's fresh palette and virtuoso handling of expressive curves were famous in the 1890s and early 1900s. Indeed, for many, Art Nouveau was *le style Mucha*.

### ALPHONSE **MUCHA**

#### 1860-1939

Mucha was born in Moravia (then part of the Austro-Hungarian Empire, now in the Czech Republic) and began his career as a stage designer at the Ringtheater in Vienna. In the 1880s, he moved to Paris, where he established himself as a printmaker and graphic artist. He produced many posters in the Art Nouveau style and branched out into jewellery design for clients such as Sarah Bernhardt. In 1922, he returned to his homeland (by then independent Czechoslovakia) and concentrated on "Slav Epic", a cycle of 20 large paintings depicting the history of the Czechs and other Slavic peoples. Mucha also designed stamps and banknotes for his home country.

# Visual tour

**KEY**

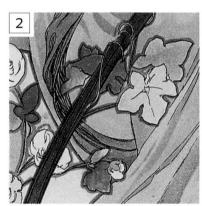

◄ **FACE AND FRAME** The young woman's face is quite simply drawn, in contrast to the detail in the hair and foliage. Her head, curving shoulders, and upper arms are embraced within a sweeping circle.

▶ **FOLIAGE** Like many of Mucha's female figures, this young woman is surrounded by swirling leaves and branches that wind around and appear to fuse with her body. This effect recalls ancient myths of metamorphosis, such as the story of the nymph Daphne, who is transformed into a laurel tree.

▲ **LETTERING** Like many Art Nouveau artists, Mucha experimented with lettering. He often rejected the traditional forms of typography in favour of expressive, flowing letters, such as this A and M with their inward-curving lines.

◄ **CURVES AND STRAIGHT LINES** The sinuous lines of the woman's dress and the twining branches are offset by the strong verticals of the panels that frame the figure.

## IN **CONTEXT**

From graphic arts to glassware, Art Nouveau was the dominant decorative style in Europe from 1890 to 1910. Widely promoted in design magazines, its organic decoration and whiplash curves were found in all areas of the decorative arts, from exclusive jewellery only affordable by the wealthy to the design of Metro stations in Paris. The style was embraced by architects including the Belgian Victor Horta and Frenchman Hector Guimard, and designers such as the American Louis Comfort Tiffany (see pp.26–27).

▲ **Paris Métro entrance designed by Guimard**

## ON **DESIGN**

In 1881, Vienna's Ringtheater burned down and Mucha lost his job as a set designer. He developed his well-known graphic style in Paris during the 1890s, producing both advertising material and decorative posters. In the poster titled Zodiac (right), originally created as a page for a calendar, Mucha's striking design features his trademark circle. He worked by making a sketch of his design, in pencil, ink, and watercolour, at full size. Here, he would have roughed out the woman's head, the outline of her hair, her jewellery, and the positions of the zodiac signs in a circle around her, then worked up the detail before the image was transferred to a the lithographic stone for printing.

▶ **Zodiac**, Alphonse Mucha, 1896

# Tiffany lamp

1900-10 ▪ LIGHTING ▪ STAINED GLASS, COPPER, AND BRONZE ▪ USA

SCALE

## CLARA DRISCOLL, LOUIS COMFORT TIFFANY

**The American artist and designer** Louis Comfort Tiffany set up his Tiffany Glass Company in New York in 1885. His elegant and colourful glassware rapidly became famous, but Tiffany's real breakthrough came when he began to produce stained-glass lampshades, in around 1898. Soon, these costly, handmade items were much sought after by fashionable decorators and homeowners on both sides of the Atlantic. The designs of the shades were influenced by the 19th-century Aesthetic Movement, and the organic flower forms and ornate decoration followed the style of Art Nouveau (see pp.24-25)

Tiffany's lampshades were constructed in a similar way to stained-glass windows in churches. Carefully cut and shaped pieces of coloured glass were arranged in abstract patterns or to form stylized images of flowers, dragonflies, and other natural subjects, then welded together with strips of copper. The lamps were painstakingly crafted in the company's workshops using Favrile glass, the name Tiffany gave to its handmade coloured glass, which had been specially formulated to create its characteristic iridescent sheen. For years, Tiffany himself, who worked closely with his artists, was thought to be the design genius behind the lamps, but the substantial contribution of unsung designer and craftswoman Clara Driscoll has now been fully recognized.

### CLARA **DRISCOLL**

#### 1861-1944

Clara Driscoll was educated at design school in Cleveland, Ohio, and at the Metropolitan Museum Art School in New York. She began work at the Tiffany Glass Company in around 1888, becoming head of the team of female glass-cutters, and staying for over 20 years. The company's records were destroyed in the 1930s, so nothing was known about Driscoll until 2005, when scholars discovered letters describing her detailed design work on Tiffany lamps. Following this revelation, Driscoll's pivotal role in creating bestselling lampshade designs, such as Dragonfly, Wisteria, and Peony, finally came to light.

A bronze cap with a small finial in metalwork acts as a strong joining point for the supporting copper strips

▲ **Full view**

## ON DESIGN

Tiffany produced many ranges of lamps and each one was made by hand, so every lamp was subtly different. Even the simplest, which had shades with geometric patterns made up of squares, ovals, and triangles, were remarkable for the diverse iridescent colours of the glass. The naturalistic lamps, with their images of flowers, butterflies, and peacock feathers, were more complex.

There were also designs moulded from a single piece of Favrile glass, the iridescent glass that Tiffany developed with the aid of Arthur J. Nash, his British-born manager. Nash added different chemicals to the glass to create colours ranging from rich blues to delicate golds and yellows, and the famous opalescent sheen. Favrile glass gave the lamps a unique quality and, because the formulae for the different colours remained a closely guarded secret, no other company was able to produce products as glisteningly elegant as those of Tiffany.

▲ **Blue lamp** with a shade and base made of Favrile glass

# "Colour is to the eye what music is to the ear"

**LOUIS COMFORT TIFFANY**

Colourful centre sections add visual interest to the flowers

Most of the illumination comes through the lower part of the shade

# Visual tour

**KEY**

▶ **RECTILINEAR DESIGN** The top third of the shade has a repeating, interlocking design that combines stems and leaves with rectilinear bands. This simplicity of form and colour palette provides a foil for the vibrant, intricate flower design beneath. Many of Tiffany's shade designs combined pictorial and geometrical patterns in this way.

1

◀ **FLOWER** Skilfully cut pieces of red glass make up the radiating flower petals. These shapes were traced onto the glass from a template, then glassworkers cut them out one by one with specialist pliers. The mottled colour and the sheen of the glass add to the richness and individuality of each lampshade.

2

▶ **LEAF** The intense red of the flowers is set off by green leaves against a golden ground. This lively palette retains a natural subtlety because of the variations in colour within each individual piece of glass. These variations were planned to give the design a handcrafted quality, whether or not the lamp was turned on.

3

▶ **BOTTOM EDGE** Most of the flowers and leaves are contained within a straight-edged, geometric border, with just the occasional leaf overlapping it. Some of Tiffany's lampshades have an irregular lower edge that follows the line of the leaves and flowers in the design.

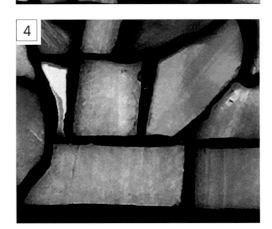

4

# Hill House chair

1903 ■ FURNITURE ■ EBONIZED WOOD ■ UK

SCALE

## CHARLES RENNIE MACKINTOSH

**One of the most strikingly original chairs** of the 20th century was created by Scottish architect-designer Charles Rennie Mackintosh for Hill House, the house near Glasgow in Scotland that he designed in 1903 for publisher Walter Blackie. Like much of Mackintosh's furniture, the chair is made of ebonized wood. Its outstanding feature is the tall ladder back that extends all the way down to the floor. Mackintosh created similar chairs for several Scottish interiors as well as for the decorative schemes he exhibited in Britain and at the Vienna Secession. He used these tall-backed chairs in two ways: as dining chairs, which helped to define an enclosed, intimate space around the table; and as stand-alone pieces, set against the walls of halls and bedrooms, where they acted as decorative, rather grand elements in the room, often echoing the architectural features of the building. The Hill House chair was specifically designed to form part of the bedroom furnishings of Walter Blackie's home. Set against the pale walls at Hill House, the strong, clean geometry of Mackintosh's design became obvious. Every part of the chair appears straight, although the ladder back rungs are curved for comfort, and the design is a formal exercise in the interplay and balance of verticals and horizontals, rectangles and squares. It recalls the bold graphics of the Vienna Secessionists, whose work Mackintosh greatly admired, as well as linear Japanese design. In marked contrast to upholstered Victorian furniture of the period, this elegant chair can also be seen as a piece of modern, abstract art that was ahead of its time.

### CHARLES RENNIE **MACKINTOSH**

#### 1868-1928

Born in Glasgow, Charles Rennie Mackintosh served an architectural apprenticeship, attended the Glasgow School of Art, and then began to work for Glasgow architects Honeyman and Keppie, initially as a draughtsman. Here he met Herbert McNair, also a draughtsman, and the pair, together with sisters Margaret and Frances Macdonald, designed and exhibited work together and became known as "The Glasgow Four". In 1996, Mackintosh won a competition to design the Glasgow School of Art, and in 1897, he was commissioned to design a chain of Glasgow tearooms. In 1900, Mackintosh caused a sensation with his interior and furniture designs for the 8th Secessionist Exhibition in Vienna. He married Margaret Macdonald and the couple occasionally collaborated on projects. Mackintosh left Glasgow in 1913, when commissions began to tail off, and spent his final years painting watercolours.

The ladder effect continues to the floor, increasing both visual impact and structural strength

The white seat forms a dramatic contrast with the dark wood

The slender legs harmonize with the other components of the chair

Closely positioned crosspieces are designed to be seen against a white wall

# Visual tour

**► GRID PATTERN**
The upper section of the chair (taking up about one quarter of the tall back section) is made up of a grid of equal-sized small squares. This pattern is characteristic of many of Mackintosh's decorating schemes.

**KEY**

1

4

2

**▲ LADDER BACK** The simple ladder back is a feature of traditional 'country' chairs in which the "rungs" are widely spaced slats of wood. In Mackintosh's version, the back is elongated and the "rungs" are square-section rather than slatted, and placed close together, to produce a geometrical effect.

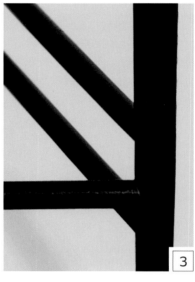

3

**▲ UPRIGHT** The ebonized timber that Mackintosh specified for this chair is reminiscent of the lacquered wood used in Japanese furniture. Slender in comparison with other Mackintosh chairs, this one was designed to stand out against the white walls of the main bedroom at Hill House.

**◄ SIDE STRETCHERS** The chair is made of timber with quite a narrow section. Mackintosh reinforced the legs with double stretchers at the sides, to provide extra strength. This double stretcher echoes the ladder-like chair back, giving the design coherence.

## ON ARCHITECTURE

In his buildings, Mackintosh combined the traditional with the modern. At Hill House, the quintessentially Scottish harled (rough-cast) exterior contains innovative interiors. The most impressive are the library and hall, with their dark wooden panelling and doors, the spacious, airy drawing room, and the elegant main bedroom. This is where the Hill House chair can be seen, set between two white wardrobes. The repeating patterns of squares on the fittings and furniture all help to unify the decorative scheme.

**▲ Main bedroom,** Hill House

## ON DESIGN

Mackintosh was successful both as an architect and a designer, and some of his most striking interiors were produced in collaboration with his wife, Margaret Macdonald, an accomplished designer of textiles, stained glass, and graphics. Mackintosh's best work was for clients such as Walter Blackie, who commissioned both exteriors and interiors. In several of Macintosh's major projects, such as the Glasgow School of Art, Kate Cranston's Glasgow tearooms, and Hill House, the furniture designs were so successful that they were manufactured and used beyond their original settings. The striking Argyle chair made for one of Kate Cranston's tearooms, its tall back surmounted by an oval, is one of the most iconic. Its contrasting curves and strong verticals make it instantly recognizable. The Willow chair, another tearoom design, has a distinctive geometrical pattern of intersecting verticals and horizontals.

**► Argyle chair,** 1897

# Flatware

1904 ▪ METALWARE ▪ ELECTROPLATED NICKEL SILVER ▪ AUSTRIA

SCALE

## JOSEF HOFFMANN AND THE WIENER WERKSTÄTTE

**This set of flatware** is almost austere in its purity and simplicity. Stylistically, it represents a total rejection of traditional 19th-century metalwork, which featured shells, scrolls, flutes, and other forms of historically inspired ornamentation. Josef Hoffmann considered such ornamentation superfluous, and preferred to concentrate on elegant lines and comfortably balanced weight.

A Moravian-born architect and designer who made his career in Vienna, Hoffmann was a key figure of the Vienna Secession and founder of the Wiener Werkstätte, two major arts and craft movements that pursued progressive design ideas. Like his fellow designers, Hoffmann was committed to applying fine craftsmanship to the production of everyday items such as chairs,

lamps, and tableware. His designs emphasized line and structure and often featured strong, geometric shapes. In this set of cutlery, which is simple and elegant in form, each piece is functional and the fine quality of the design is clear. The bowls of the spoons swell generously, but the heads of the forks are almost the same width as their handles – both are designed to sit comfortably in the hand as well as to look good on the table.

### WIENER WERKSTÄTTE

Josef Hoffmann (1870-1956) and Kolomon Moser founded the Wiener Werkstätte (Vienna Workshops) in 1903. Their aim was to produce simple, high-quality items for the home, in the craft tradition of designers such as William Morris (see p.17). The designs were strongly influenced by the ideals of the Vienna Secession. Initially, the emphasis was on rectilinear designs and simple, abstract patterns, but after about 1910, they produced more ornate designs to appeal to a wider market. By 1905, the Werkstätte employed around 100 workers, making a range of items from metalwork to furniture, and opened shops in cities including New York. The workshops eventually closed in 1932 because of difficulties caused by the Great Depression.

▲ **Werkstätte founder** Josef Hoffmann stands (far left) with Kolomon Moser (far right), artist Gustav Klimt (standing third from left), and other members of the Vienna Secession in this group portrait.

The fish knife lacks the conventional broad, swelling blade

**Dessert fork**    **Fork**    **Fish knife**    **Butter knife**    **Knife**

# Visual tour

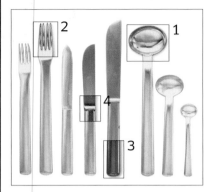

**KEY**

➤ **SPOON BOWL** The deep oval spoon bowls, which join seamlessly to their handles, are elliptical. Unusually, the ellipse is wider than it is deep, making it easy to sip from the short side of the bowl.

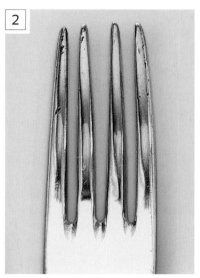

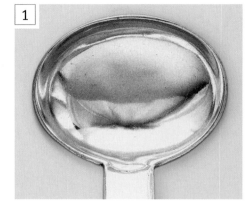

➤ **FORK TINES** Both the forks in the place setting have simple, narrow heads that are only slightly wider than their handles. The long, elegant tines, positioned quite close to each other give the forks a clean, modern look.

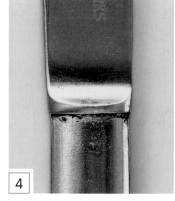

▲ **KNIFE JOINT** A slight bulge marks the point where the knife blade joins the handle. This masks and strengthens the joint between the handle and the steel knife blade. It also protects the user, stopping the hand from sliding down the handle to the blade.

◄ **BASE OF HANDLE** Whereas many traditional knives and forks have a decorative flourish at the end of the handle, the Hoffmann flatware is perfectly plain. The slight chamfer looks elegant and is very comfortable to hold.

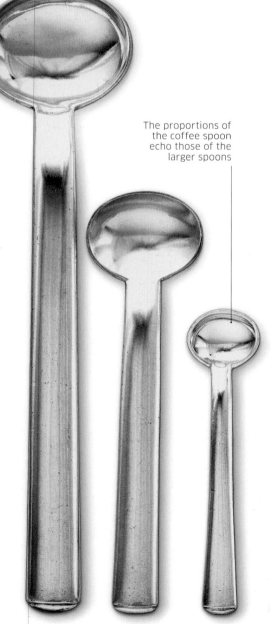

The proportions of the coffee spoon echo those of the larger spoons

**Soup spoon**　　**Teaspoon**　　**Coffee spoon**

## IN CONTEXT

In 1897, a group of Viennese artists broke away from the Viennese Academy in protest against the academy's old-fashioned outlook and, in particular, its refusal to allow foreign artists to display work at its exhibitions. The breakaway movement paralleled others in Munich and Berlin and was known as the Vienna Secession. It was led by painter Gustav Klimt and included designers and architects such as Josef Hoffmann, Kolomon Moser, and J.M. Olbrich. They built their own headquarters and exhibition hall in Vienna, held exhibitions, and published a journal called *Ver Sacrum* (Sacred Spring), which acted as a conduit for their work and ideas, as well as illustrating products made by the Wiener Werkstätte. The Secessionists wanted to create a union between the fine arts and the applied or decorative arts and their work ranged from Klimt's decorative paintings to Moser's bold graphics. Their fresh and inclusive approach to art influenced many of Austria's architects and artists.

◄ **Secession Building, Vienna** Designed by J.M. Olbrich and built in 1898, this was the headquarters of the Vienna Secession. Compared with the traditional architecture of the city, the building's strong, rectilinear form, the patterned ornament of the façade, and the golden openwork globe on top were both radical and modern.

# Electric kettle

1908 ▪ PRODUCT DESIGN ▪ NICKEL-PLATED BRASS AND CANE ▪ GERMANY

SCALE

## PETER BEHRENS

**In 1908, the German AEG company** introduced the first successful electric kettle, an appliance that heralded the idea of the new all-electric kitchen. This classic design, produced by the versatile Peter Behrens, featured a metal body, removable lid, heatproof handle, and electric socket. It set the style for electric kettles throughout most of the 20th century.

The kettle had its origins in a practical problem. When domestic electricity became widespread around 1900, demand peaked in the mornings and evenings, but there was a surplus of power during the day. Companies like AEG, keen to exploit this unused daytime supply, began to produce electrical appliances aimed at the consumer. The electric kettle fitted the bill, encouraging consumers to boil water at any time of day. It was small, inexpensive, and easy to use, and the design combined modern detailing with traditional elements, such as the overall shape and lid, to reassure customers who were hesitant about the new technology.

Behrens was one of the pioneers of modern industrial design, especially in the home. To quote his own words: "Design is not about decorating functional forms – it is about creating forms that accord with the character of the object and that show new technologies to advantage."

The turned knob is similar to those on traditional pots and pans

The body is made of nickel-plated brass

Solid metal base

## PETER **BEHRENS**

### 1868-1940

German architect and designer Peter Behrens trained as a painter and was influenced by Jugendstil (Art Nouveau) and German Secessionist groups. In 1900, he was invited to join an artistic colony at Darmstadt, where he designed and built his own house. In 1907, he became one of the founders of the Deutscher Werkbund, a group set up to improve the quality of industrial design in Germany. The same year, AEG invited Behrens to advise the company on its corporate identity and expanded the brief to cover all aspects of design within the company. Behrens' influence was huge and Walter Gropius, Ludwig Mies van der Rohe (see p.58), and Le Corbusier (see pp.56-7) all worked under him early in their careers.

# Visual tour

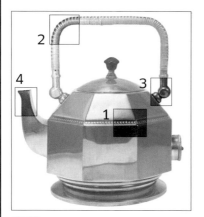

**KEY**

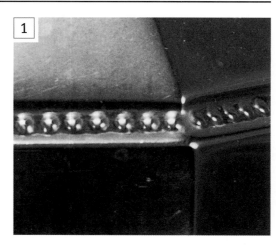

> ➤ **BODY** The simple, octagonal shape of the kettle's body is relieved by discreet ornament in the form of a band of beading, a touch that recalls the coffee pots and tea kettles of an earlier era. This and other details gave the design a touch of familiarity that appealed to householders who were not yet accustomed to electrical appliances.

1

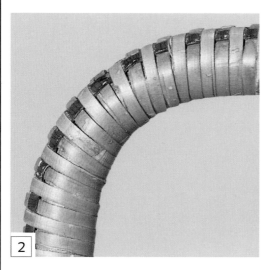

2

3

4

▲ **HANDLE** The handle is wrapped with woven cane, which is a poor conductor of heat, to protect the user's hand. It also gives the kettle a hand-crafted appearance. The height and shape of the handle were carefully considered from a practical perspective. By holding the corner of the handle, users could tilt the kettle to pour water.

▲ **JOINT** A metal joint makes the link between the sloping face of the kettle body and the vertical section of the handle. This functional component, which appears to be derived from engineering, illustrates how the design incorporates utilitarian details.

▲ **SPOUT** The spout has a traditional, gentle curve and lip so that it pours effectively. In cross-section, however, its angular form matches the octagonal body. Even in this small detail, the design reveals a mixture of the modern and the traditional.

## ON **DESIGN**

When AEG appointed Peter Behrens as their architect and design consultant, they were breaking with tradition. Never before had a major company hired one person to advise on every aspect of their design work, from the graphics of their logo to the architecture of their factories. Behrens rose to the challenge, designing a range of high-quality, functional products, including fans, kettles, and clocks. Both designer and employers were well aware of the importance of marketing and branding, and Behrens designed logos and advertisements for the company (right). At first, he displayed the company name in flowing and ornate lettering, before settling on the simpler serif capitals within a plain border that the company still uses today. Under Behrens, AEG could rightly claim that design was at the heart of everything it produced.

**1907**

**1908**

**1908**

**1912**

# Red Blue chair

C.1917 ■ FURNITURE ■ PAINTED BEECHWOOD AND PLYWOOD ■ NETHERLANDS

SCALE

## GERRIT RIETVELD

**Few objects seem to sum up a moment** in design history more effectively than Gerrit Rietveld's Red Blue chair. With a plywood seat and back that look like free-floating planes, and a framework of uprights and horizontals, it is widely regarded as an icon of de Stijl (the Style), the avant-garde art and design movement that brought Modernism to the Netherlands in the aftermath of World War I – at a time when people were looking forward to rebuilding a new Europe.

The chair's geometry and primary colours epitomize the pure, harmonious forms advocated by de Stijl's followers and reflect the work of the group's most famous painter, Piet Mondrian. The original chair, however, was made from unstained wood and the colours were only applied later, probably in the early 1920s. In breaking away from past styles, Rietveld had created a pared-down and totally unfamiliar chair, yet he also regarded it as a practical design. He made it from standard cuts of timber and hoped that it would eventually be mass-produced, but the chair was never produced on a large scale. Some people found its stark lines uncompromising; others complained that the flat wooden seat was simply not comfortable. When Rietveld was told this, he did not disagree, but simply stated, "It's not really a chair: it's a manifesto". As a plea for a fresh way of looking at furniture design, it is still eloquent.

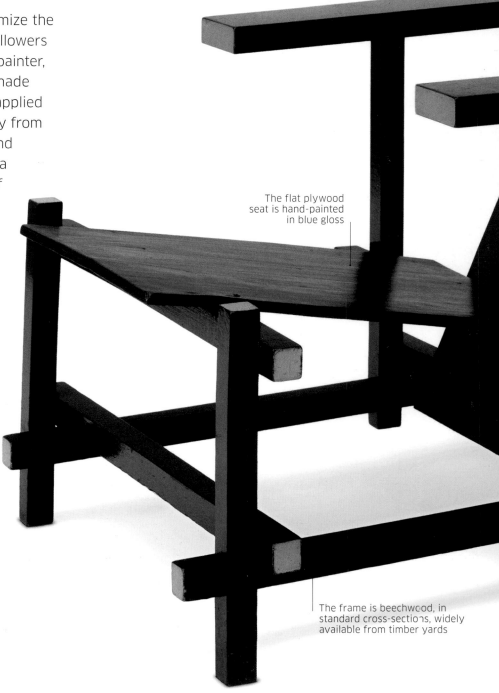

The flat plywood seat is hand-painted in blue gloss

The frame is beechwood, in standard cross-sections, widely available from timber yards

---

### GERRIT **RIETVELD**

#### 1888-1964

Dutch architect and designer Gerrit Rietveld started work in his father's furniture workshop at the age of 12. He then became a jeweller's draughtsman and took classes with the architect P.J.C. Klaarhamer. In 1917, he started his own furniture workshop and two years later began to contribute to the de Stijl journal, which brought his work to a wider audience. As well as designing furniture, he worked on designs for typography and building projects such as the Schröder House (see opposite) in Utrecht. His architectural work included one-off houses for individual clients, designs for mass-produced "housing modules", and prefabricated buildings made from steel frames and concrete panels. Later in his career, he took on high-profile projects, such as displays and pavilions in international exhibitions, including the Dutch pavilion for the 1954 Venice Biennale. As his reputation grew, he was given commissions for prestigious buildings, such as the Gerrit Rietveld Academie and the Van Gogh Museum in Amsterdam. The latter was completed by his associates after his death.

# Visual tour

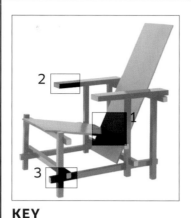

**KEY**

◀ **SEAT AND BACK**
These two panels, which do not touch each other, are made of painted plywood and are set at angles that are designed for comfort. Both the seat and back are attached to their respective supports by hidden brackets and screws, and seem to float independently of the chair frame.

The back is set at an angle of 30°, which Rietveld considered ideal for relaxation

▲ **ARMREST** The squared-off ends of the arms and stretchers are highlighted in bright yellow. The painted ends seem to indicate that the lines extend infinitely. On first impression, the chair looks like a series of squares and rectangles, similar in effect, coincidentally, to the geometric shapes that characterize Piet Mondrian's paintings.

▲ **JOINT** Most of the joints occur where the sides of pieces of timber meet, so there is no need for the complex joints traditionally used in furniture when two pieces of end-grain meet at a corner. The black stretchers, legs, and crosspieces of the chair are skilfully fixed together with cylindrical wooden pegs that cannot be seen.

## ON **ARCHITECTURE**

De Stijl promoted art that exploited the harmonious relation of planes, straight vertical and horizontal lines, and a palette consisting of white, black, and primary colours. The innovative house in Utrecht that Rietveld designed (below), is true to these ideals. The surfaces of the planes are white or grey, the linear elements are in primary colours, and the

whole building is made up of right angles – even the windows only open at 90 degrees to the façades. The basic shape of the building is that of a cube, broken at many points by protruding floor panels and supports. The large windows and interior sliding partitions also open up the building, creating an adaptable interior that is flooded with light.

The leg terminates simply, without any ornament or foot

◀ **Schröder House,** Utrecht, Netherlands. Rietveld designed this modest-sized house in 1924, in collaboration with his client, Truss Schröder-Schräder. The interior floors, walls, and furnishings echo the clean lines and colours of the exterior.

1920–1929

- **L'Élégance perfume bottle** René Lalique

- **Silver pitcher** Johan Rohde

- **AGA cooker** Gustaf Dalén

- **Dobrolet poster** Alexandr Rodchenko, Varvara Stepanova

- **Bauhaus poster** Joost Schmidt

- **E1027 occasional table** Eileen Gray

- **Cité armchair** Jean Prouvé

- **Nord Express poster** A.M. Cassandre

- **LC4 chaise longue** Le Corbusier, Charlotte Perriand, Pierre Jeanneret

- **Barcelona chair** Ludwig Mies van der Rohe

# L'Élégance perfume bottle

1920 ▪ GLASSWARE ▪ FROSTED GLASS ▪ FRANCE

SCALE

## RENÉ LALIQUE

**The name René Lalique** became identified with sophisticated glassware in the first half of the 20th century, and his perfume bottles, like this exquisite example in frosted glass for D'Orsay's L'Élégance fragrance, typify his style. Lalique created all kinds of glass objects – vases, jewellery, light fittings, statuettes, and even clocks – but established his reputation with the bottles that he made for several different perfumiers. He used ordinary glass rather than crystal and adopted industrial processes such as blowing glass into moulds, developing production techniques to match the delicacy of his designs. He formed the moulds precisely using wax models and finished the glass immaculately to give it an opalescent sheen, or added colours to highlights or backgrounds.

Lalique's designs were inspired by animal and plant forms or were taken from classical subjects, like the figures on the L'Élégance bottle, but his interpretation of such motifs

was distinctive. In this mould-blown rectangular bottle in frosted glass, the couple, draped in diaphanous cloth, are shown in low relief, and Lalique picked out the background with a sepia tint. Both the figures and the angular bottle anticipate the elegance of the early Art Deco style.

The stopper has horizontal ribs

The sinuous shapes of the figures contrast with the rectilinear form of the bottle

### RENÉ **LALIQUE**

#### 1860-1945

The great French glass-maker René Lalique started out as a jewellery designer, working for many prestigious Parisian jewellers in the 1880s and 1890s. Some of his jewellery incorporated inexpensive materials, such as enamel and glass.
In 1908, perfumier François Coty commissioned Lalique to design his labels and in 1909, Lalique opened a glassworks east of Paris, where he designed and produced bottles for Coty and experimented with new kinds of decorative glassware. In the 1920s and 1930s, as well as working on interior designs and other projects, he produced a vast amount of glassware, from exclusive one-off vases and statuettes to mass-produced, but still high-quality, items. Lalique's display at the 1925 Paris International Exhibition (see opposite) was a triumph, and he became a leading exponent of Art Deco.

# Visual tour

**KEY**

▶ **STOPPER** Lalique's perfume bottles also had stoppers made of glass. This one is tinted with the same sepia tone as the bottle itself. It is domed in shape with horizontal ribs to help the user grip it, and contrasts with the strong, rectangular shape of the bottle.

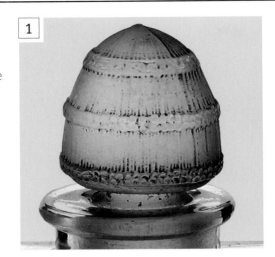

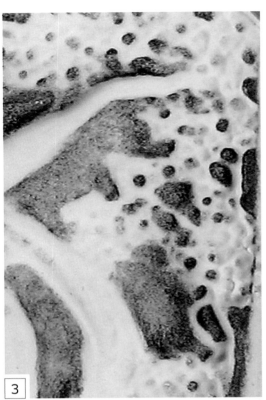

▶ **TWO FACES** With their faces positioned centrally, just beneath the neck of the bottle, the couple seem to have come together briefly during their dance. This sophisticated image hints at the romantic encounter implicit in the purchase of perfumes.

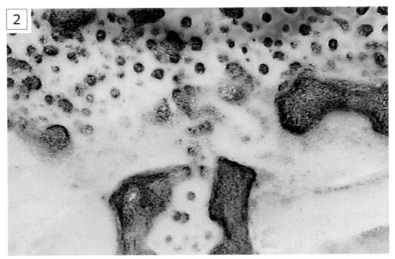

▲ **FILIGREE WORK** The stylized foliage that surrounds the dancing couple and is used as a framing device is picked out in white and sepia, which has been applied in the spaces between the foliage and branches with great precision.

## ON DESIGN

Lalique's use of industrial production techniques enabled him to produce a huge variety of glassware including car mascots, clock cases, items for the dressing table, lamps, and a large number of vases and perfume bottles. Paperweights and car mascots gave him the scope to produce striking animal figures including Antelope (below left). Many of the vases and bottles, such as Dolphin (below) and Telline (below right), were created in clear or opalescent glass, with decorative details picked out in low relief. Coloured stains were then incorporated to add richness to the surface of the glass. Sometimes Lalique varied the colour sequence, first selecting a glass coloured in deep red or amber and then adding a white tint to make the whole effect more sculptural.

▲ Antelope paperweight

▲ Dolphin vase

▲ Telline perfume bottle

## IN CONTEXT

In 1925, the Exposition Internationale des Arts Décoratifs et Industriels Modernes was held in Paris. Decorative items on show, with distinctive motifs such as lightning flashes and sunbursts, typified the new 1920s style. Embraced by Hollywood and influencing designs from jewellery to architecture, this style later became known as Art Deco, after the exhibition.

▲ **Art Deco interior**, designed by Mme B.J. Klotz

# Silver pitcher

1920 ■ METALWARE ■ SILVER AND WOOD ■ DENMARK

SCALE

The angled neck extends down to join the handle

A gentle curve defines the form of the body

## JOHAN ROHDE

**The great Danish silversmith Georg Jensen** opened his workshop in Copenhagen in 1904, after having been apprenticed as a goldsmith at 14 and then trained as a sculptor. His early designs were influenced by Art Nouveau style (see pp.24–25), and rooted in the fine workmanship of the Arts and Crafts movement (see p.16–19), but as his business grew he began to employ designers with new ideas. One such designer was Johan Rohde, whose designs were sleeker and more stylized than Jensen's. One of Rohde's most impressive pieces is this silver pitcher, which he designed in 1920. It is a striking example of the flair and simplicity shown by Danish design of the period.

Rohde's pitcher draws its shape from the organic curves of Art Nouveau, but has none of the style's characteristic decoration of flowers and foliage – it is an exercise in pure form. It is also a practical vessel, with a swelling, flat-bottomed body that is very stable, a handle that fits the hand comfortably, and a lip that pours with ease. The pitcher does not seem to belong to any particular style – Rohde had moved beyond Art Nouveau but had not yet embraced the more angular forms of Art Deco or Modernism – yet it embodies the perfect combination of form and function that remained the holy grail of the great modern designers.

### JOHAN **ROHDE**

#### 1856–1935

Danish designer, sculptor, metalworker, and architect Johan Rohde trained as a doctor before switching to fine art and teaching anatomy to artists. In 1904–05, he designed silverware for his own use and commissioned Georg Jensen to make it. Jensen was impressed with the design, and by 1906 Rohde was creating items for Jensen to manufacture and sell. He produced several designs for Jensen, including a range of flatware called *Konge* (Acorn), which proved to be very popular. In addition, Rohde created furniture, textile, and metalwork designs.

# Visual tour

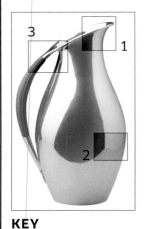

**KEY**

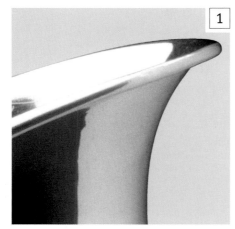

◄ **POURING LIP AND RIM**
The lip seems to grow organically out of the angled top of the pitcher in a transition so subtle that it is impossible to tell where one part ends and the other begins. The body of the vessel also curves seamlessly into the rim.

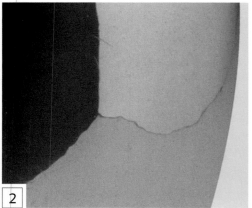

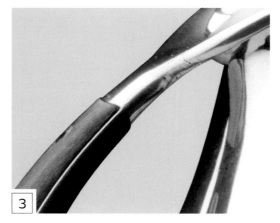

▲ **SURFACE** The pitcher's polished surface, as flawless as a Georgian coffee pot, is designed to look prestigious, but also to look at home in any context. Whether placed next to antique tableware or Modernist cutlery, the reflective surface and unadorned form provide a harmonious accent.

▲ **HANDLE** The wooden handle is not necessary for insulation, as it would be on a tea or coffee pot, but is more practical than a silver handle, which would pick up the user's fingermarks. It is integrated smoothly into the silver supports, in another of the vessel's long, fluid curves.

## ON **DESIGN**

Although Georg Jensen liked to give his designers a free hand, encouraging them to be creative in order to produce their best work, he also knew his market. Reflecting on the purity of the most modern designs, he realized that the public needed time to learn to appreciate them. Jensen therefore encouraged his designers to incorporate more traditional elements into some designs, such as naturalistic ornament, or to draw on the shapes of earlier vessels to add touches of familiarity.

▲ **Cosmos water pitcher, 1925** The design of this jug by Rohde, with its upswept, embellished handle and ornamented body, takes a more traditional form than his celebrated silver pitcher.

## IN **CONTEXT**

Georg Jensen and other Danish designers responded to changing fashions, but by avoiding slavish imitation they created their own versions of the evolving styles of the 19th and 20th centuries. Under the influence of Rohde and others, the ornate Art Nouveau of 1900 (as in Bing and Grondal's pear-shaped vase, below left) was replaced by the pared-down style of Harald Nielsen's candleholder (below), in which a swirl of decoration enlivens a simple underlying form. By the 1920s, designers such as Sigvard Bernadotte embraced Art Deco, using geometrical patterns such as the diapers on the cocktail shaker (below), or natural motifs, such as the birds in Arno Malinowski's brooch, within a rectangular frame (below right).

▲ **Pear-shaped vase**
Bing and Grondal, 1902

▲ **Candleholder**
Harald Nielsen, 1915

▲ **Cocktail shaker**
Sigvard Bernadotte, 1920s

▲ **Brooch**
Arno Malinowski, 1930s

# AGA cooker

1922 ■ PRODUCT DESIGN ■ CAST IRON WITH ENAMEL FINISH ■ SWEDEN

## GUSTAF DALÉN

SCALE

**The AGA cooker has become a symbol** of rural domesticity in the UK since its invention in 1922 by Swedish engineer and scientist Gustaf Dalén, and its enamelled, cast-iron form has hardly changed in appearance in that time. Dalén designed the AGA following a failed experiment that robbed him of his sight. While recuperating at home, he created an energy-efficient appliance to lighten the domestic load that his disability placed on his busy wife.

Simpler to operate than many European ranges of the period, the AGA was intended to be kept on at all times during the cold winter months, so that it heated the house, warmed water, and dried laundry, as well as functioning as a stove. The ovens and hotplates were arranged so that their distance from the burner regulated their temperature, making one oven ideal for roasting and baking, and the other for simmering and warming. The AGA cooked using radiant heat, which helped to retain moisture and flavour.

Manufactured from heavy cast iron in the UK since 1929, the AGA was built to last decades. Gas, electricity, and oil-fired models are now made, and new technology enables the user to control when the AGA cooker is on or off.

### GUSTAF **DALÉN**

**1869-1937**

Engineer Gustaf Dalén was managing director of the Swedish Gas Accumulator Ltd company, where he worked with acetylene gas. He invented the sun-valve, a device that made a light come on at dusk and turn off at dawn, so that lighthouses could work automatically. Dalén was awarded the Nobel prize for physics in 1912. He also devised a way of storing acetylene that made it safe to use for welding, but lost his sight testing acetylene cylinders. Although blind, Dalén continued to lead the company from 1913 until his death.

# Visual tour

**KEY**

▶ **DOOR AND HINGE** The AGA cooker's hinges are cast as part of the door. Their horizontal lines emerge seamlessly from the door's surface, fitting into rings on the cooker's body. The door is thicker in the middle, where it covers the opening of the oven, and thinner towards the edges, where less insulation is necessary.

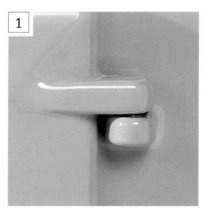

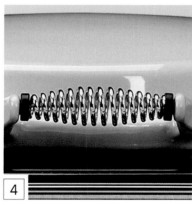

▶ **AGA NAMEPLATE**
The AGA brand name is displayed prominently on the front of the oven, its cursive script set within an oval panel. The three letters stand for the name of Dalén's company, the Svenska Aktiebolaget Gas Accumulator. The AGA cooker is now made exclusively in the UK. The trademark is registered to AGA Rangemaster plc.

▲ **HOTPLATE COVER** The heavy, domed metal covers are hinged at the rear and opened using the coiled metal handles at the front. These covers retain the heat, so whenever the AGA is on, the hotplates are ready for use.

◀ **CATCH TO OVEN DOOR** The door catch has a very simple design. The door casting has a knob that fits into a hook on the body. To open the door, the user simply lifts it slightly to free the protrusion from the hook.

▼ **Facsimile of a 1940s AGA**

On the original models,
a thermostat controlled
heat from the burner;
here it is maintained as
a decorative feature

Handrail

The display shows
the temperature
in the oven

The hotplates are
a generous size to
accommodate more
than one pan

# Dobrolet poster

1923 ▪ GRAPHICS ▪ COLOUR LITHOGRAPH ON PAPER ▪ USSR

SCALE

## ALEXANDR RODCHENKO, VARVARA STEPANOVA

**Alexandr Rodchenko and Varvara Stepanova** were at the cutting edge of design in the USSR in the 1920s. When Dobrolet airline launched a campaign for investment after Lenin had allowed limited private enterprise in the Soviet state, the pair came up with the most up-to-date graphics. Their arresting poster, with its intersecting forms and dynamic angles, followed the principles of Constructivism, a radical art movement that embraced modern technology and promoted art as a commodity in the service of society.

Across the middle of the poster is the image of a Junkers aircraft, the plane that made up the bulk of Dobrolet's fleet, set at a diagonal to convey a sense of upward movement. The rest of the design consists of slogans in bold, modern letter forms, insisting that anyone who is not an investor in Dobrolet is not a true citizen of the Soviet Union. Graphics such as these proved influential far outside the USSR, inspiring Western designers to experiment with bold, sans serif letter forms.

# Visual tour

**KEY**

◄ **ANGULAR LETTER FORMS**
The form of Cyrillic script used in the poster is deliberately angular – circles and ellipses appear as rectangles, curves as straight lines. The clarity and modernity of the poster's typography suggest that the airline itself is both efficient and modern, offering a state-of-the-art service for its passengers and investors.

▲ **DIAGONALS** The ascending movement of the aircraft establishes a series of diagonals – the wing, fuselage, tailplane, and propeller – that give the composition a sense of dynamism. The diagonals are mostly depicted in the plain white colour of the paper and stand out clearly against the flat expanses of green and red.

◄ **ARROW AND PROPELLER**
The Constructivists often used arrows in their posters, to draw the viewer in and involve him or her directly in the artwork. This arrowhead forms the tip of the wide red band, a framing device that runs around the poster, ending at the propeller, the focus of the image.

► **EXCLAMATION MARK**
Russian graphic artists liked to use large exclamation marks, which act as calls to attention and signal that the poster contains an important message. Visually this device forms one side of the bold red rectangle that frames the image.

## RODCHENKO, STEPANOVA

### 1891–1956, 1894–1958

In 1910, the painter, designer, and photographer Alexandr Rodchenko attended the Kazan School of Art in Odessa. Here he met fellow-artist Varvara Stepanova, who later became his wife. Before the 1917 Revolution, the pair were already prominent avant-garde artists. After 1917, they concentrated their energies on "art for the people": Rodchenko designed graphics for posters, books, and films; Stepanova focused on graphics, textiles, and set design. Rodchenko was also involved in reorganizing art schools and museums in the post-revolutionary period.

## IN CONTEXT

The Constructivist movement began in Russia after the 1917 Revolution. The Constructivists were influenced in part by earlier Russian abstract art, with its strong sense of structure and geometry. In the wake of the Revolution, artists took a rational, pragmatic approach, using art to help the construction of a socialist society. The movement was most influential in the applied arts – graphics, architecture, theatre design, fashion, and film. Some of the Constructivists concentrated on practical projects, designing clothes, and domestic items such as stoves. Among the Constructivists' most notable achievements were Vladimir Tatlin's designs for workers' clothing and his unbuilt Monument to the Third International (1920); the graphics of Rodchenko and El Lissitsky; and the textile designs of Stepanova and another talented female artist, Liubov Popova. The movement flourished in the USSR until the early 1930s, when Stalin demanded that artists turn to socialist realism. Constructivist works were, however, still widely exhibited in the West, and continued to influence the spread of Modernism, inspiring designers at the Bauhaus in Germany (see p.46).

▲ **Vladimir Tatlin** stands beside a model of his Monument to the Third International, known as Tatlin's Tower, 1919.

# Bauhaus poster

1923 ■ GRAPHICS ■ COLOUR LITHOGRAPH ON PAPER ■ GERMANY

SCALE

## JOOST SCHMIDT

**► ORIGINAL POSTER**
The personal symbol signet of sculptor and typographer Joost Schmidt is at the top right of the original poster, perfectly aligned with the composition.

**Designed to promote the first Bauhaus exhibition**, in 1923, Joost Schmidt's poster caused an immediate sensation with its bold, geometric forms and dramatic, block-like lettering. Strong and innovative, it sent out a powerful visual statement about the identity of the Bauhaus, a radical school of art and design founded in Weimar by the architect Walter Gropius in 1919.

Under Gropius's direction, the Bauhaus aimed to bring together art and technology. Gropius called the school the Staatliches Bauhaus, which literally means "State House of Construction". Based on the Arts and Crafts (see p.19) idea of a guild that would combine architecture, sculpture, and painting, the Bauhaus developed a strict, craft-based curriculum that would turn out designers capable of creating useful and beautiful everyday objects. As time went by, the aims of the school shifted to stress the importance of creating work suitable for mass production, and this was reflected in their designs, which favoured functionality and simplicity over the use of ornament.

The Bauhaus came into conflict with the German authorities from the start and, to justify its existence, was ordered in 1923 to mount an exhibition of its accomplishments. Schmidt's poster demonstrates all the early characteristics of the "New Typography" developed by the Bauhaus to establish its corporate identity. The text, made up of modern fonts, is inserted into an abstract composition of geometric shapes so that the most important pieces of information – the words Bauhaus, exhibition (*Ausstellung*), and Weimar – catch the eye.

The poster had already been printed when the dates of the exhibition were changed. Schmidt found an elegant solution to this problem by creating two labels with the new exhibition dates. These were printed on red and white paper and simply pasted onto the finished posters.

The original Bauhaus did not last for long. The school moved from Weimar to Dessau, and then to Berlin, but was finally forced to close down by the Nazi Party in 1933. Despite its brief life, however, it has proved to be perhaps one of the most influential design schools ever.

## THE **BAUHAUS**

### 1919-33

The Bauhaus was not just an art and design school, but a collective of talented and influential designers. It was founded by architect Walter Gropius who was succeeded as director, in 1928, by Hannes Meyer, and, in 1930, by Ludwig Mies van der Rohe (see pp.58-59), both of whom were also architects. Many of the Bauhaus teachers, such as Paul Klee and Wassily Kandinsky, were celebrated artists. Painter and photographer László Moholy-Nagy joined the school in 1923 and established a new direction, with a greater emphasis on mass production.

Marcel Breuer (see p.59), who ran the cabinet-making workshop from 1924 to 1928, revolutionized furniture design with chairs made from tubular steel, while weaver Gunta Stölzl ran the textile workshop, creating innovative, abstract fabrics. In the metalwork studio, Marianne Brandt, Wilhelm Wagenfeld, and Christian Dell created many successful design prototypes. After the Bauhaus closed, many of the key figures emigrated to the United States, where they influenced generations of young architects and designers.

**► Bauhaus designers** Marianne Brandt, Christian Dell, László Moholy-Nagy, Hans Przyrembel, Wilhelm Wagenfeld, and others

# Visual tour

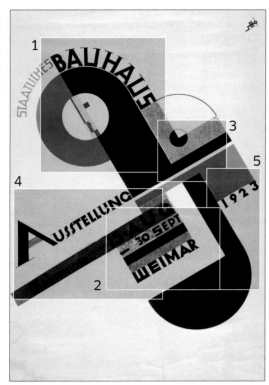

**KEY**

1

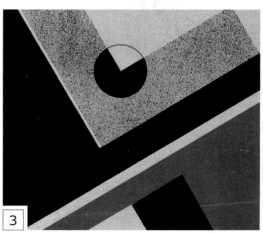

3

▼ **EXHIBITION DATES** When the 1923 exhibition was postponed, Schmidt created red and white labels with the revised dates, which he then personally pasted onto all the printed posters. The diagonal red bar with the new dates, 15th August–30th September, creates a strong visual anchor for the overall design.

▲ **INTEGRATED IMAGES AND TYPOGRAPHY** The Bauhaus designers were the first to combine images and typography in this way. The words Staatliches Bauhaus (State House of Construction), the name of the school, are wrapped around the official Bauhaus symbol created by artist Oskar Schlemmer, which is placed inside the red and black circle, forming a focal point.

2

▲ **GEOMETRIC FORMS** Simplified geometric shapes – the circle, square, and triangle – were a defining feature of Bauhaus design. Students were taught that, "the clear geometric form is the one that is most easily comprehended... every possible form lies dormant in these formal elements".

**4**

▲ **STRONG DIAGONALS** The abstract composition of diagonal bars and circular segments makes the key information, such as the word *Ausstellung* (exhibition), jump out. The striking diagonals create impact and are typical of Bauhaus.

**5**

▲ **COLOUR CONTRASTS** Schmidt restricted his use of colour to black, red, and yellow. Together with the stark, geometric shapes, this makes the composition visually striking and instantly recognizable. Bauhaus students had to study colour theory as part of their foundation course, and the school's promotional material used a distinctive range of colours and shapes, giving it a strong corporate identity.

## IN **CONTEXT**

At the Bauhaus, typography was seen both as a "tool of communication" and as a form of artistic expression. The typography workshop experimented with new fonts, creating bold, block-like lettering that became symbolic of the school's identity. In 1925, Herbert Bayer created Universal, a simple, geometric font (see below). Not only were there no serifs (tails at the end of each stroke); he did not use capital letters either. The Bauhaus typography, with its characters stripped of all ornamentation, initially caused an uproar among critics, but had a lasting influence.

abcdefghi
jklmnopqr
stuvwxyz d

HERBERT BAYER: Abb. 1. Alfabet
„g" und „k" sind noch als
unfertig zu betrachten

Beispiel eines Zeichens
in größerem Maßstab
Präzise optische Wirkung

sturm blond

Abb. 2. Anwendung

399

## ON **DESIGN**

The Bauhaus set up the first ever art, craft, and design foundation course, which introduced new students to the school's goals and encouraged them to experiment in a wide range of disciplines. Following their immersion in Bauhaus theory, students entered specialized workshops, which included metalworking, cabinet-making, weaving, pottery, and typography (see above). Although the initial aim of the Bauhaus was to unify the arts through craft, the focus soon changed to developing design prototypes suitable for mass production, and the school adopted the slogan "Art into Industry".

All the Bauhaus workshops stripped design down to its bare bones, exploring geometric forms and experimenting with new materials, such as tubular steel and plywood, to create simple, functional pieces. All surface decoration was completely eliminated. The cabinet-making workshop created minimalist furniture, using industrial materials, while the metalwork studio created beautiful, modern items, such as lights and tableware, and the textile workshop created abstract textiles for use in the school's buildings.

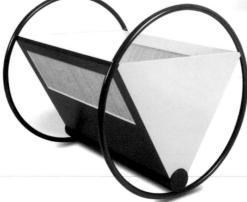

▶ **Bauhaus cradle** Designed by Peter Keler in 1922, this brightly-coloured cradle was inspired by the paintings of the Bauhaus tutor Wassily Kandinsky.

▲ **Table lamp** Wilhelm Wagenfeld's 1924 lamp is made up of simple geometrical forms in metal and glass. The round shade was of a type previously used in industrial lighting.

# E1027 occasional table

1927 ▪ FURNITURE ▪ CHROME-PLATED TUBULAR STEEL AND GLASS ▪ FRANCE

SCALE

## EILEEN GRAY

**When Irish-born furniture designer** and architect Eileen Gray collaborated with her partner Jean Badovici in the design of their innovative holiday home at Roquebrune-Cap-Martin in the South of France, she also created all the furniture and fittings. Impressed by the chairs Marcel Breuer (see p.59) and his colleagues at the Bauhaus were producing with tubular steel, Gray used chrome-plated tubular steel for several pieces, including a small table designed for the guest room. This table, which is beautiful in its simplicity, consists of just a few lengths of tubular steel that have been joined together seamlessly and chrome plated to give them a shiny surface. The top is a glass disc and its circular steel frame is mirrored by a broken circle of steel tubing that forms the table's base. These two round elements are linked by a pair of steel uprights, one of them piercing the glass top, and the whole table is finished immaculately.

This combination of steel and glass produces a table that looks very light and insubstantial, but which is strong and heavy enough to be functional. Gray had a very specific function in mind. She designed it for one of her sisters, who liked to have breakfast in her room. For this purpose she set the twin steel uprights that support the top to one side so the table top could be pulled over the bed. Gray's house was modest in scale and the table, with its extendable steel uprights for raising or lowering the height of the glass top, was a practical and adaptable piece of furniture for the space. At a low setting, the piece doubled as an occasional table. Gray named the table "E1027" after the Roquebrune house, a somewhat anonymous name for a home, but the letter and numbers form a personal code: Gray's initials (E for Eileen, plus 7 for the 7th letter of the alphabet, G, for Gray), enclose those of Jean Badovici (J and B, the 10th and 2nd letters of the alphabet).

# Visual tour

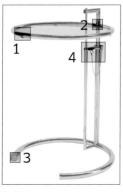

**KEY**

▶ **EDGE OF TABLE** The table top is made of crystal glass and is supported by small brackets under the steel frame. The tubing protrudes above the level of the glass, preventing items on the table from sliding off the edge.

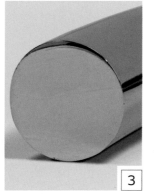

▶ **STEEL JOIN** One of the most impressive aspects of the table is the flawless finish. Where the main upright intersects with the top's circle of tubular steel, the quality of the join is superb and appears almost seamless.

▲ **ADJUSTING MECHANISM** To make the table secure at different height settings, Gray used a simple pin inserted into holes in the upright. This is a very basic, low-tech solution that works well. Aware that a loose pin could easily be mislaid, the designer added a metal chain to attach the pin to the table.

◀ **BASE** The tubular steel elements of the table are capped at the ends so that they look more like metal bars than hollow tubes. This creates a neat finish and also gives the table an appearance of solidity and strength.

The glass top can be raised by pulling up the cross-piece

## EILEEN **GRAY**

**1878–1976**

After studying fine art at the Slade School in London, Scottish-Irish Eileen Gray trained in the art of lacquer in Paris. She produced interior schemes and furniture that were initially decorative but became increasingly Modernist in style. Encouraged by her partner, Romanian architect and writer Jean Badovici, she turned to architecture in the 1920s, designing houses for herself, a studio for Badovici, and Modernist furniture for their interiors. Her work was fashionable in the 1930s, attracting the attention of Le Corbusier and others, but after World War II (when most of her drawings and belongings were lost), she was no longer in the public eye and her work was largely forgotten. A resurgence of interest in the 1970s encouraged manufacturers to revive some of her Modernist designs, including the E1027 table.

"To create, one must first question everything"

**EILEEN GRAY**

All the tubes are finished in polished chrome

The broken circle of the base makes it possible to position the table around the leg of a bed or sofa

## ON **DESIGN**

Although she was a contemporary of Modernist designers such as Breuer and Le Corbusier (see p.57), Eileen Gray worked outside the usual networks of modern design. She was critical of some of the Modernist ideas, asserting that a house was not just a "machine", as Le Corbusier famously remarked, but should satisfy the heart as well as the mind. Gray worked out her ideas on her own – and sometimes in collaboration with Jean Badovici – in numerous pencil drawings. She developed the concept of a small house in which spaces were flexible and pieces of furniture, such as the E1027 table or a cupboard that turns into a writing desk, could be adapted to different uses.

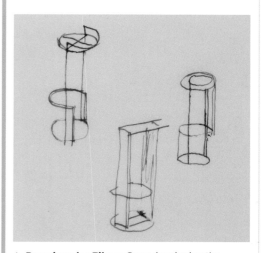

▲ **Drawings by Eileen Gray** developing the concept of an adjustable-height table

## IN **CONTEXT**

Gray learned how to apply lacquer decoration to wood from a Japanese craftsman working in Paris. Although this was a very traditional craft, Gray developed ways of using it in a modern manner. The outstanding examples of this were her screens made of lacquered wood panels joined by metal rods. By building up a screen out of a number of panels, Gray turned a piece of furniture into an abstract sculpture that could be adjusted to create different shapes. She had broken up the traditional form of the screen in an almost Cubist fashion.

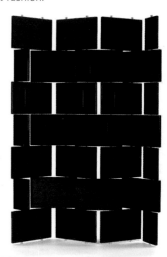

▶ **Screen** made of lacquered wood and metal rods, 1922

# Cité armchair

1927 ▪ FURNITURE ▪ FOLDED STEEL, FABRIC, AND LEATHER ▪ FRANCE

## JEAN PROUVÉ

SCALE

**In 1927, French designer and manufacturer** Jean Prouvé produced a new armchair for student residences in his home town of Nancy. Prouvé had trained as a blacksmith, but also knew about the latest methods for folding and bending steel, and he used these techniques in this striking design. With its strong, folded-steel frame supporting a fabric-covered seat and arms made of leather straps, the chair was designed to stand up to years of heavy use. The Cité armchair incorporated a beautifully balanced series of straight lines and curves with an emphasis on function and an absence of applied decoration, and soon found design-conscious buyers outside the university.

Prouvé's Modernist designs took a different direction from the creations of his contemporaries, such as Ludwig Mies van der Rohe (see p.58) and Le Corbusier (see p.57). Their furniture, although minimalist in form, shows off the richness of materials such as polished steel and leather upholstery and was designed to harmonize with Modernist interiors. Prouvé, on the other hand, concentrated on functionality, robustness, and ease of manufacture. These

interests give the Cité armchair a robust, austere quality and toughness that won Prouvé commissions to create furniture for hospitals, government offices, and schools. With innovative designs such as this chair, Prouvé succeeded in bringing Modernism to the masses.

### JEAN **PROUVÉ**

**1901-84**

The son of a founding member of the Art Nouveau School of Nancy, Jean Prouvé rejected that ornate style. The success of his functional designs brought him commissions for hard-wearing furniture, and he became a manufacturer on a large scale. Prouvé had a social conscience and used his knowledge of steel to build prefabricated houses, including homes for refugees. He also became a leading producer of curtain-wall façades for tall buildings. For many years, Prouvé gave hugely popular lectures to students at the Conservatoire des Arts et Métiers in Paris.

# Visual tour

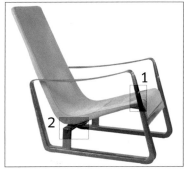

**KEY**

▶ **UPRIGHT** The chair frame is made of a strip of sheet steel that has been folded lengthways so that instead of being flat it has an inner channel. This channel gives the steel strength and creates a space for the leather strap that forms the chair arm. Crosspieces prevent the strap from slipping out of the channel.

1

2

▲ **SEAT SUPPORT** Prouvé noticed that chairs often break at the point where the legs join the seat, and knew that the original student users of his Cité chairs were likely to treat them roughly. The robust support of the Cité chair consists of a heavy piece of steel that is attached firmly to the base.

## ON **DESIGN**

Although Prouvé's designs were usually elegant, appearance was only one of the factors he considered when working on a new product. His main interests were function and, above all, construction – he famously said, "Never design anything that cannot be made". Many of his drawings survive, and these show him working out how to join and combine components – for example, the joint between the side frame and the crosspiece on the Cité chair, and the way the leather strap runs through the channel in the chair frame. Prouvé's drawings have great clarity and simplicity, and he used them to analyse every step of the production process, to describe how to produce a particular piece, and, later in his life, to explain his methods to design students – creating beautifully executed details in chalk on lecture theatre blackboards.

▶ **Working drawing**, Jean Prouvé, c.1948

High, supportive fabric back

A leather strap runs around the entire frame

The line of the chair arm echoes that of the seat

The steel frame is folded to make a three-sided channel

The classic finish was lacquered black; red and beige were also available

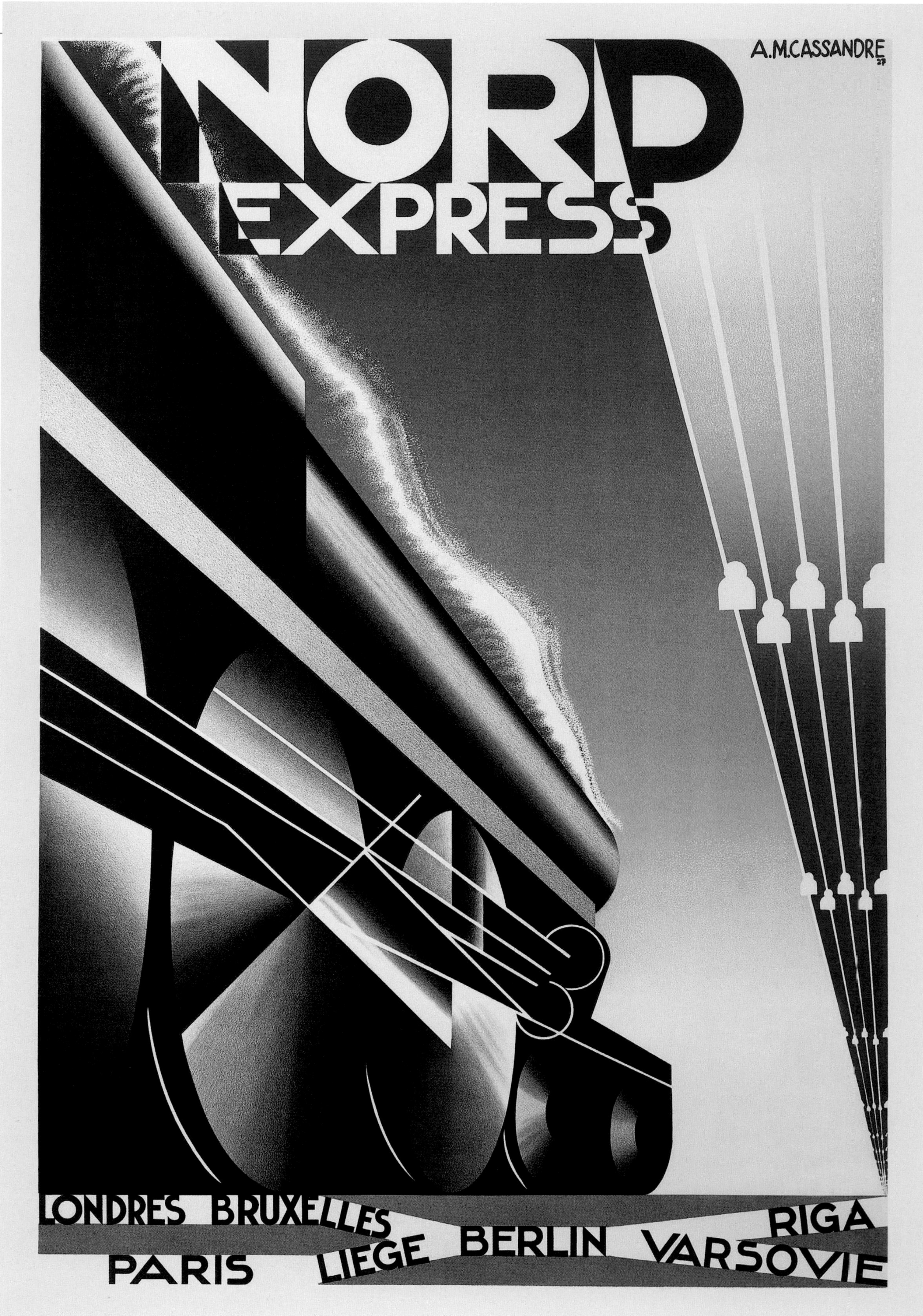
A.M.CASSANDRE 27
NORD
EXPRESS
LONDRES BRUXELLES
PARIS
LIEGE
BERLIN
VARSOVIE
RIGA

# Nord Express poster

1927 ■ GRAPHICS ■ COLOUR LITHOGRAPH ON PAPER ■ FRANCE

SCALE

## A.M. CASSANDRE

In the 1920s, the artist A.M. Cassandre transformed the appearance of the streets of Paris with a series of radical advertising posters. The most dramatic, stylized images were those for the Compagnie des Wagons-Lits, which operated night trains across Europe, such as this poster for the Nord Express route. Choosing a low, track-side viewpoint to emphasize the locomotive's great bulk and power, Cassandre makes the engine appear to tower above the viewer. A skilful use of perspective creates the illusion of headlong speed and is achieved by a series of dynamic diagonals – the engine's boiler, its connecting rods, the plume of steam, and the telegraph wires, all running towards the vanishing point, set in the bottom right corner. The engine itself is drawn as a series of shaded circles and planes, in a style that shows the influence of Cubist painters such as Pablo Picasso and Juan Gris. By combining these effects with strong lettering, Cassandre established a new style of advertising art that he also applied to posters for shipping lines, wine-makers, and other clients. His bold, modern style inspired artists and designers after World War I to produce more daring graphic imagery.

### A.M. CASSANDRE

#### 1901–1968

Adolf Jean-Marie Mouron studied art in Paris. He adopted the name A.M. Cassandre in 1922, when he began designing posters. Influenced by both Cubism and Surrealism, he was soon in demand as a poster artist and designer of typefaces. His success continued, with many commissions for French and foreign firms, until World War II, when he served in the French army. In the post-war period he worked as a set designer and created a logo for Yves Saint Laurent, but he suffered from depression and committed suicide in 1968.

# Visual tour

**KEY**

▶ **DESTINATIONS** At the bottom of the poster, Cassandre represents the train journey in the form of a stylized route diagram. The pale strips representing the track, with the names of key destinations in capital letters, stand out against the grey background. The lines converge at Berlin then fork for Riga and Warsaw (Varsovie).

1

▶ **DISPLAY LETTERING** The name of the train service is displayed in very large letters. Cassandre's were some of the first posters big enough to be legible from a moving vehicle. The letters have striking red highlights where they intersect with the diagonals – a clever device to catch the viewer's eye.

2

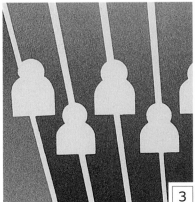

3

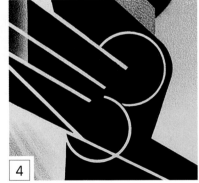

4

▲ **ENGINE** The stylized treatment of the wheels and connecting rods is striking. The wheel is broken up in an almost Cubist way and the connecting rods are sketched in as white lines – to suggest detail that would be invisible when the train was travelling at speed.

◀ **TELEGRAPH INSULATORS** The white shapes represent the ceramic insulators used on telegraph wires and were a common sight along railway lines. The steep, overhead perspective suggests the speed of the train as it hurtles towards its destination.

# LC4 chaise longue

1928 ■ FURNITURE ■ CHROME-PLATED TUBULAR STEEL AND LEATHER ■ FRANCE

SCALE

## LE CORBUSIER, CHARLOTTE PERRIAND, PIERRE JEANNERET

**Le Corbusier was an architect first and foremost**, and his Pavillon de l'Esprit Nouveau at the 1925 Paris Exhibition was a prototype of sparsely furnished and standardized living. He described the house as a "machine for living in" and called furniture "domestic equipment". For the Maison la Roche in Paris, he commissioned Charlotte Perriand and Pierre Jeanneret to design chairs suitable for "conversation, relaxing and sleeping". The streamlined LC4, inspired by the 18th-century chaise longue, fulfilled the last of these needs and became an icon of Modernist furniture. Le Corbusier described it as the "true resting machine", and its exposed framework did indeed make it look like a piece of equipment. The materials used were radical. Perriand combined tubular steel – previously reserved for industrial products – with sleek black leather. The LC4 not only looked good – it was also designed to be supremely comfortable. Cassina is the only company authorized to produce this chaise longue since 1964.

Fully adjustable cushion

The support is tilted slightly to accommodate the frame

The steel frame balances on the base, and can be easily adjusted

### CHARLOTTE **PERRIAND**

**1887-1965**

One of the most influential furniture designers of early Modernism, Charlotte Perriand helped to introduce the "machine age" aesthetic to interiors. Born in Paris, she joined Le Corbusier's studio after making a name for herself designing tubular-steel furniture. The collaboration lasted for ten years. Perriand worked in Japan and Vietnam from 1940 to 1946, returning to France to work with designer Jean Prouvé (see p.52), and contributing to Le Corbusier's Unité d'Habitation in Marseille. She later worked on furnishings for Méribel ski resort, as well as the League of Nations building in Geneva, and Air France offices in London, Paris, and Tokyo.

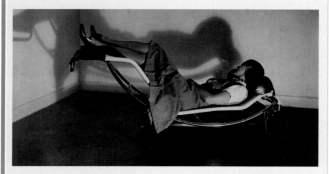

▲ **Charlotte Perriand** poses on an LC4

# "Chairs are architecture, sofas are bourgeois"

**LE CORBUSIER**

# Visual tour

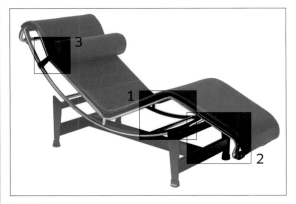

## KEY

**► EXPOSED FRAMEWORK**
Traditionally, the structure of large pieces of furniture was hidden by upholstery. In keeping with the Modernists' machine aesthetic, the framework of the LC4 is on full display. The tubular-steel frame and black steel legs are integral to the design.

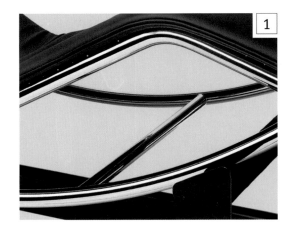

**◄ TUBULAR STEEL** Sweeping lengths of chrome-plated tubular steel form the seat and the cradle beneath it, giving the chaise longue its graceful shape. The strength and pliability of tubular steel, combined with its light weight, made it the ideal material for Modernist furniture, and so did its associations with engineering and industrial mass production. Tubular steel is both functional and minimalist – qualities that held particular appeal for Le Corbusier and his colleagues.

**◄ HEADREST** The leather-covered headrest is a comfortable bolster shape and can be adjusted as required. Such attention to detail shows that the chaise longue was designed with comfort in mind. The base of the chair is adjustable too, making the LC4 one of the first pieces of truly ergonomic furniture. The chair was suitable even for those with back problems.

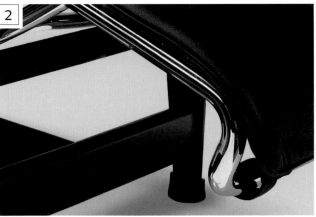

The leather-covered seat rests on concealed springs

## ON **ARCHITECTURE**

One of the most most influential architects of the 20th century, Le Corbusier was born Charles-Edouard Jeanneret-Gris at La Chaux de Fonds, Switzerland. He trained as an artist before travelling extensively, then settled in Paris where he wrote his radical book *Vers une Architecture* (Towards an Architecture) and later adopted the name Le Corbusier. From 1922, he worked with his cousin, Pierre Jeanneret, and began to design houses. Le Corbusier's work changed dramatically over the years, from the regional vernacular of his early houses in Switzerland to his Modernist villas of the 1920s, such as the celebrated Villa Savoye (below), which featured "free façades" of non-supporting walls and spacious open-plan interiors. Later work ranged from the brutalist housing of the Unité d'Habitation at Marseille (1945–52) to the sculptural style exemplified by the chapel at Ronchamp (1950–55).

▲ **Villa Savoye**, Poissy, France, 1928–31

# Barcelona chair

1929 ▪ FURNITURE ▪ CHROME-PLATED STEEL AND LEATHER ▪ GERMANY

SCALE

## LUDWIG MIES VAN DER ROHE

**The Barcelona chair**, designed by German architect Ludwig Mies van der Rohe, is one of the most recognizable pieces of furniture ever produced. Its combination of clean lines, minimal frame, and luxurious leather upholstery have made it a favourite as a stylish easy chair to furnish interiors ranging from elegant reception rooms to the lobbies of prestigious office blocks. The chair still looks modern, even though it was designed in 1929 for the German Pavilion that Mies built at the Barcelona International Exhibition, where a pair of the chairs were used as thrones for the King and Queen of Spain.

For the frame of the Barcelona chair, strong, chrome-plated steel bars were welded together in a curved X shape. Two of these modified Xs were then joined together with crossbars, which support a series of leather straps. Deep, rectangular, leather-covered cushions for the seat and backrest sit on top of the straps. Using a similar, flattened version of the X frame, Mies also

produced leather-cushioned stools, and occasional tables topped with sheets of glass 2.5cm (one inch) thick. The furniture was displayed to perfection against the marble and glass walls and chrome-plated columns of the Barcelona Pavilion. The genius of Mies, however, lay not only in his creation of a piece of furniture that reduced the chair to its essentials, but also in creating a design of such timeless elegance that the chair would enhance any modern setting.

Leather straps are attached to the steel crossbar

The steel frame splays out gracefully at the end to form the foot

---

### LUDWIG MIES VAN DER ROHE

#### 1886–1969

Born in Aachen, Germany, Mies van der Rohe worked in his father's stone-carving workshop and attended a trade school. He was apprenticed on several building projects before moving to Berlin. There, he worked with architect Peter Behrens and then set up on his own, designing a string of key Modernist buildings, including the Tugendhat House in Brno, then Czechoslovakia. Most of these conform to his famous maxim, "Less is more", and the furniture he designed for his buildings became popular in its own right. In 1930, Mies was appointed the last director of the Bauhaus, then moved to the USA, where famous works include New York's Seagram Building.

The leather cushion is a traditional tufted button style

A minimum of the framework is visible under the cushion

# Visual tour

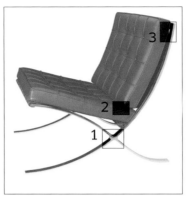

**KEY**

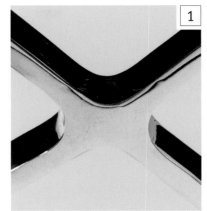

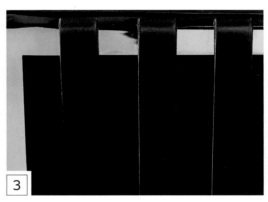

◀ **FRAME JOINT** A substantial weld joins the bars at the crossing point, giving the chair considerable strength. The frame is made of steel that is beautifully finished and generously proportioned. Mies wanted the chair to be heavy – when he placed a chair at the ideal point in a room, he wanted to discourage users from moving it around.

▲ **CUSHION CORNER** The piping at the edges of the leather cushions creates a clean line that emphasizes their rectangular shape. Although the upholstery is deep, as befits a comfortable easy chair, the rectilinear outline is not compromised, and the Barcelona chair fits in with the geometry of the architect's Modernist interiors.

▲ **LEATHER STRAPPING** Mies specified leather straps to support the seat and back of the chair. The leather straps harmonize with the leather cushions and are placed close together to give good support. They also maintain the appearance of luxury and quality, enhancing the chair's purpose to be seen as a sculpture in the round.

## IN **CONTEXT**

Mies was exposed to the most up-to-date thinking about design and architecture through his early training with Peter Behrens and his experience at the Bauhaus (see p.46). Many Modernist designers produced furniture by combining traditional materials, such as canvas and leather, with engineered products, such as tubular steel. Marcel Breuer (see p.46), who studied at the Bauhaus, was one of the first to use tubular steel for a chair frame – it is said that he got the idea from the handlebars of his bicycle. The result was the B3 chair (also known as the Wassily chair, in honour of the painter Wassily Kandinsky, who taught at the Bauhaus). The curving steel frame complements the tight canvas of the arms, seat, and back.

▲ **B3 Wassily chair, Marcel Breuer, 1925**

- **Moka Express coffee-maker** Alfonso Bialetti

- **London Underground map** Harry Beck

- **Ericsson telephone DHB 1001** Jean Heiberg,
Christian Bjerknes

- **Pencil sharpener** Raymond Loewy

- **Anglepoise lamp** George Carwardine

- **Ekco radio AD65** Wells Coates

- **Savoy vase** Alvar Aalto

- **Kodak Bantam Special** Walter Dorwin Teague

- **Volkswagen Beetle Model 1300** Ferdinand Porsche

- **Knuten candelabra** Josef Frank

- **B.K.F. chair** Antonio Bonet, Juan Kurchan,
Jorge Ferrari Hardoy

1930 – 1939

# Moka Express coffee-maker

C.1930 ■ PRODUCT DESIGN ■ ALUMINIUM AND PLASTIC ■ ITALY

SCALE

## ALFONSO BIALETTI

**In the early years of the 20th century**, Italian café owners scored a hit with espresso – strong coffee produced very rapidly using a machine that forced almost-boiling water under pressure through ground coffee beans. At first, espresso machines were large, expensive, and confined to cafés, but Italian metalworker Alfonso Bialetti developed a device that was compact and simple enough to produce espresso at home. His Moka Express coffee-maker changed the game.

Working with engineer Luigi de Ponti, Bialetti based his design on the kind of silver or silver-plated coffee pots often seen in well-to-do Italian homes, but chose an octagonal shape that harmonized with the Art Deco style fashionable in the 1930s. His design was made from aluminium, which kept the cost down, and contained an ingenious arrangement of parts that forced water up through the grounds when the coffee-maker was heated on the stove. Bialetti is said to have came up with the idea after watching Italian women launder linen in a tub with a central pipe that drew up hot water from the bottom over the washing. The Moka Express works in a similar way. Although sales were modest to begin with, Bialetti's coffee-maker proved a huge success when it was aggressively marketed after World War II. Today, 90 per cent of Italian households own one, and they are sold all over the world.

Heatproof plastic, as on this model, replaced the rubber used on early models

The connecting hinge was added at a late stage in the design process

A screw connection links the upper and lower sections

The sturdy base withstands heat and pressure

### ALFONSO **BIALETTI**

**1888–1970**

Born in the Piedmont area of northern Italy, Alfonso Bialetti was a metalworker in France for ten years. After World War I, he returned to Piedmont and set up his own workshop. The idea for the Moka Express first came to him in the 1920s, but it took years of modifications before he finally began to sell the coffee-maker in 1933. He built up sales locally and by the beginning of World War II, he was making about 10,000 units per year. After the war, Bialetti's son Renato took over the business and increased production dramatically to around 1,000 coffee-makers per day.

# Visual tour

**KEY**

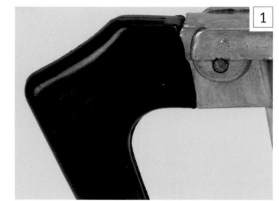

**◄ HANDLE** The handle of the Moka Express was carefully designed to feel comfortable in the hand. The angled upper surface acts as a thumb rest, while the fingers curl around the contoured inner part, which provides a firmer grip and prevents them from coming into contact with the hot metal.

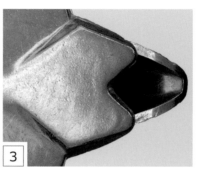

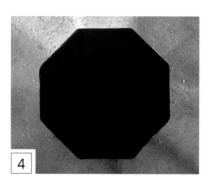

**▲ BASE SECTION** The octagonal form of the base is a variation on traditional coffee pots seen on middle-class tables all over Europe. Bialetti went through a long process of trial and error to get the size and thickness of the sides just right, so that the pot would withstand the high pressure of the boiling water inside.

**▲ SPOUT** This takes a simple form, with triangular sides and a small pouring lip. The hinged top covers part of the spout and is designed to keep in the heat – a useful feature for people who do not wish to drink all their coffee immediately.

**▲ KNOB** The eight-sided knob on the lid echoes the shape of the body, and its angular form matches those of the handle and the triangular spout. The styling of the coffee pot is recognizably Art Deco, while also reflecting earlier traditions.

## ON DESIGN

Bialetti designed the Moka Express to be easy to use and straightforward to clean, making full use of aluminium technology. In essence, the coffee maker consists of a lower vessel that holds the water, an upper section from which the coffee is poured, and a filter between the two. The two main elements – the water tank and top section – unscrew, revealing the filter and funnel. The user takes these two sections apart, fills the tank with water up to the level of the safety valve, and installs the filter funnel. This is filled with coarsely ground coffee, the parts reassembled, and the whole unit placed on the hob. When the water heats, it is pushed up the stem of the funnel, into the ground coffee, and through the filter. A gurgling sound indicates that the espresso is ready to drink.

Only one modification was made to the Moka Express after Bialetti's death: the addition of a logo in the form of a caricature of its maker.

Hinged lid

Filter

Gasket

Filter funnel

Safety valve

Water vessel

## ON PRODUCTION

One of the skills Alfonso Bialetti had learned when working in France was how to cast aluminium objects in reusable cast-iron moulds. He realised that this technique would be ideal for producing strong, but light coffee-makers. The lower section of each Moka Express was cast by hand, ensuring a robust coffee-maker with a good finish. Today, Moka Express coffee makers are still produced in exactly the same way, and they are still made of aluminium.

**▲ Casting** base sections of the Moka Express.

# London Underground map

1931 ■ GRAPHICS ■ PRINT ON PAPER/CARD ■ UK

## HARRY BECK

SCALE

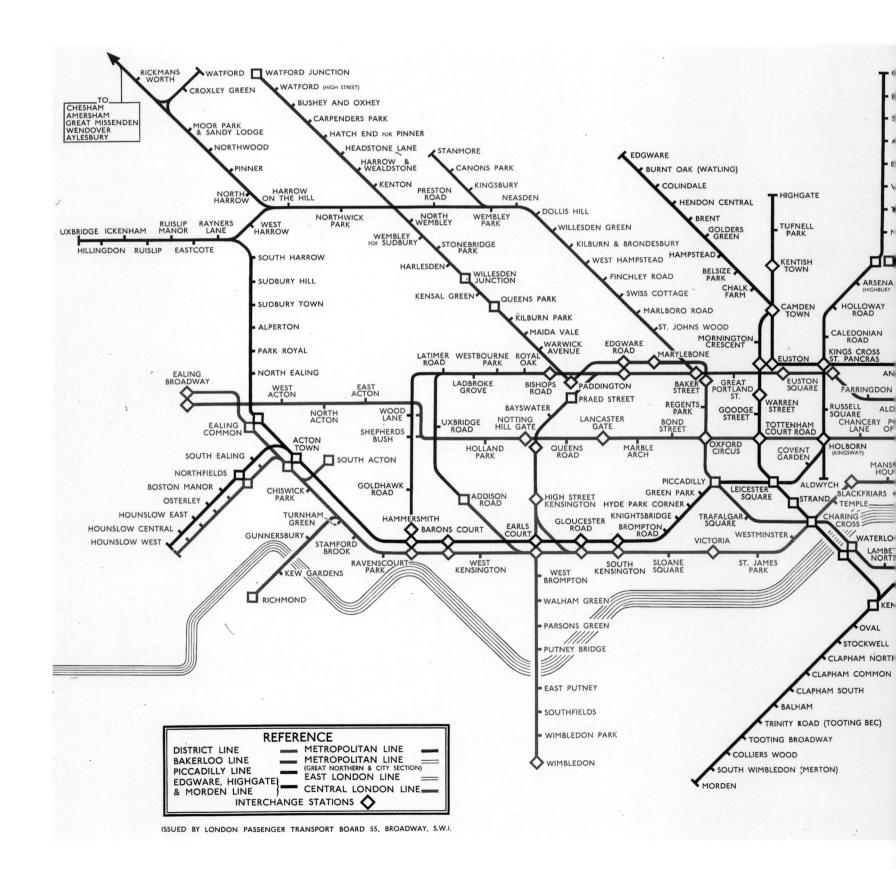

**By the early 1930s, London** had a complex underground railway network with several lines serving a large part of the city. Passengers guided themselves around it using a complicated map with snaking lines and a crowded central section. In some versions, the ground-level streets were also shown. The system was not easy to navigate and, in 1931, a young engineering draughtsman, Harry Beck, came up with the idea of making the map much simpler to use. He departed from the convention of plotting exact physical locations and drawing them to scale, and created a radical, diagrammatic map that showed how the stations and lines related to each other. Beck's map depicted each line in a different colour, and the lines on the map ran only horizontally, vertically, or at 45-degree diagonals. Most stations were indicated by a short coloured line, except for the interchange stations, which were represented by a diamond shape, so that passengers could see easily where to change from one line to another. The only ground-level feature on the map was the River Thames.

When Beck presented this map, his superiors were sceptical and thought it would simply confuse passengers. In 1933, however, they were persuaded to test it out, and produced a small number of pocket fold-out maps. When these were handed out at stations, the clarity and legibility of the layout made them instantly popular. Soon, larger versions were printed for station walls. The new map became part of the network's 1930s Modernist corporate identity, which was masterminded by managing director Frank Pick. In spite of later alterations by other hands, today's Underground map is remarkably similar to Beck's original, and inner-city transit companies all over the world have mapped their own systems in a similar way. Beck's map is seen as one of the great achievements of 20th-century design and its functional approach is regarded as typical of the modern movement. Beck, however, knew little about modern art: he simply created his map with the same logic and care that he employed to draw an electrical circuit diagram.

> # "I like maps that try to simplify a network of roads or railroads (as the London Underground schemes do it)..."
>
> **OTTO NEURATH**

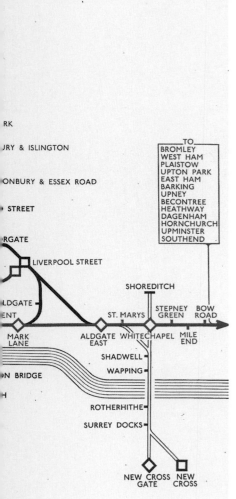

## HARRY **BECK**

### 1902–74

Henry "Harry" Charles Beck was employed as an engineering draughtsman at the London Underground signals office. Realizing that the existing route map was becoming cluttered as new lines and stations were added, he created an improved version in his spare time. Beck left London Transport in 1947, but made revisions to the map until 1960, when the work was assigned to another designer. Beck also created maps for the overground rail network around London and for the Paris Métro, but these were not adopted. In later life, he taught typography and design at the London College of Printing.

# Visual tour

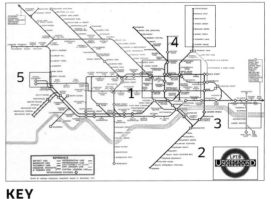

**KEY**

▼ **COLOUR CODING** Beck represented each of the lines that made up the London Underground system in a different colour, as on the earlier maps. These lines had originally been run by different train companies, but the operators gradually merged and the London Passenger Transport Board took over control of all of them in 1933. Beck's map was published in the same year and the new layout with its colour-coded lines helped the board establish the identity of the Underground system.

▼ **ROUNDEL** The roundel symbol, with its circle and horizontal bar, is displayed prominently on the map. It was first used in 1908, when the Underground lines adopted a symbol consisting of a blue band across a solid red disc. The band carried a station name or the word "Underground". In 1917, Edward Johnston, who had also created the network's typeface, redesigned the symbol with a red circle instead of the solid disc. In subsequent years the roundel was used widely on stations and in publicity material..

**1**

▲ **RIVER THAMES** London's river, represented by parallel thin blue lines, is the only non-railway feature on the map. Its inclusion anchors the map in the geography of the capital and it is a familiar landmark by which Londoners orient themselves. Passengers can see at a glance which stations are north of the river and which are to the south.

COCKFOSTERS

ENFIELD WEST

SOUTHGATE

ARNOS GROVE

BOUNDS GREEN

WOOD GREEN

TURNPIKE LANE

MANOR HOUSE

4

▲ **UNDER CONSTRUCTION** In the 1930s, the Underground network was still growing  Beck developed the idea of including parts of the system that were under construction as broken lines on the map. All the stations were named in the sans-serif typeface that Edward Johnston had designed specially for the network in 1916.

# EALING BROADWAY

5

▲ **INTERCHANGE STATIONS** On earlier maps, it was not always obvious whether two lines were meeting at an interchange, or simply passing close to one another. Beck's map made these details much clearer. He depicted interchange stations as diamonds, and the two adjacent colours indicated which lines served each interchange.

## ON **DESIGN**

Although many of London's Underground lines curve their way around to avoid subterranean obstacles, Beck decided that it would be clearer to illustrate them diagrammatically, as straight lines, with a few standardized curves to indicate changes of direction. He developed the idea in a series of drawings, allowing space in one corner for a key, and marking the places where lines went under the Thames. Beck then refined the diagram at points where several lines joined or crossed, working out the clearest way to show these complex intersections. A similar process took place whenever the network was altered and Beck revised the map.

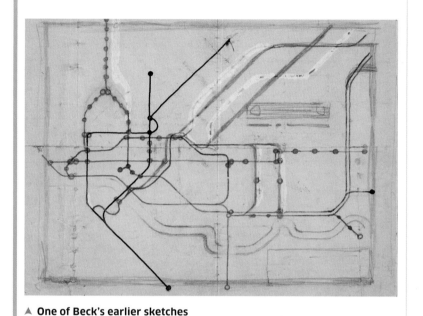

▲ **One of Beck's earlier sketches**

## IN **CONTEXT**

Beck's diagrammatic approach to transport mapping proved influential, with other cities adopting a similar approach involving straight lines, standardized angles, colour coding, and the inclusion of few above-ground features. From Madrid to Seoul, the underground rail maps of many capital cities recognizably derive from Beck's work in the 1930s. Although the details may vary, with designers choosing different systems of colour coding and different symbols for stations, travellers quickly grasp the visual language of these maps when they visit an unfamiliar city. In many countries, similar maps have been used to show other kinds of transport networks, especially overground railways.

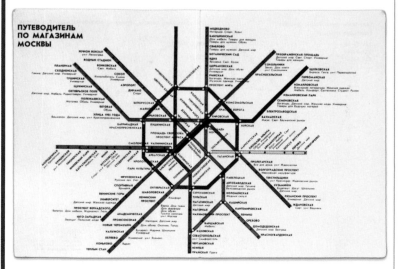

▲ **Moscow metro map** The metro system in Moscow, with its radial lines, like spokes, and inner circle section, lends itself well to diagrammatic mapping.

# Ericsson telephone DHB 1001

1931 ▪ PRODUCT DESIGN ▪ BAKELITE AND METAL ▪ SWEDEN/NORWAY

SCALE

## JEAN HEIBERG, CHRISTIAN BJERKNES

**In 1931, the Swedish telecommunications company** Ericsson and the Norwegian Electrisk Bureau launched a new telephone that set the standard for home and office phones for the next 40 years. With its bold, sculptural shape, front dial, and handset resting in a cradle, it had everything that people wanted in a telephone at the time, in a robust form that was made to last.

The Ericsson telephone represents a successful marriage of art and technology. Norwegian engineer Christian Bjerknes designed the components and included an internal bell at a time when many house phones relied on a separate bell mounted elsewhere. He also developed devices to reduce interference, leading to better sound quality in the earpiece. Artist Jean Heiberg, who was primarily a painter, designed the body and handset. He approached the telephone as if it were a piece of sculpture and found the perfect material in Bakelite, which could be moulded to the exact shape he had in mind. The body of the telephone was carefully modelled to accommodate the bell, inner workings, and dial, and its cradle was integrated with the main body rather than formed as a separate component. The smooth, curved handset was easy to pick up and hold and the earpiece fitted comfortably against the head. Simple to operate and good-looking, with clean, modern lines, the Ericsson telephone achieved rapid success and was used, imitated, and produced in modified versions all over the world.

### JEAN **HEIBERG**

#### 1884-1976

Norwegian painter Jean Heiberg studied art in Oslo, Munich, and Paris, where he studied at the academy of Henri Matisse. He returned home to Norway and painted in a style influenced by his French teacher, as well as producing sculptures in bronze and designing a few pieces of furniture. In 1935, he became professor at the Norwegian National Academy of Fine Arts, a post he held for 20 years, with an interruption during the Nazi occupation of Norway, when he ran an underground art academy with fellow artists. Heiberg was a major influence on 20th-century Norwegian artists.

# Visual tour

**KEY**

1

▲ **NUMBERS AND DIAL** The dial rotor is made of hard-wearing metal and the numbers are engraved on the centre disc. Letters could be added in the spaces behind the finger holes if the local telephone system used them. Later versions of the telephone have a Bakelite dial.

◄ **CORNER MOULDING** Heiberg designed the telephone with crisp edges, giving it a purposeful look. This combination of hard edges and curves was easy to produce in Bakelite – it only took about seven minutes to make the body of the phone.

► **MOUTHPIECE** The mouthpiece has a curved extension, making it easy for the user to hold it close to the mouth. This feature helped to exclude external noise and made reception clearer for the person who was listening at the other end.

# "The Swedish type of telephone"

## ON **DESIGN**

In the 1920s, most telephones were made out of sheet metal and wood and the mouthpiece and earpiece were often two separate parts, which the user held in either hand. By using Bakelite, Jean Heiberg was able to combine these two parts in a single handset. He developed his design using plaster models, refining details such as the plinth-like base, which some people thought he had based on the platform that supports the columns of an ancient Greek temple. He also tried out a double ring of moulding around the earpiece and mouthpiece, but this detail was dropped from the final design.

▲ **Plaster model** of the Ericsson telephone body by Jean Heiberg

The Bakelite handset has a gently curved shape

The integral cradle is robust, which reduces breakages

The lower part of the body contains the bell

This panel is designed to hold a card displaying the telephone number

The feet make the telephone stable

# Pencil sharpener

1931 ▪ PRODUCT DESIGN ▪ CHROME-PLATED STEEL ▪ USA

SCALE

## RAYMOND LOEWY

**The celebrated promoter of streamlined** design, which he applied to everything from cars to cocktail shakers, Raymond Loewy became known as "the man who reshaped America". Streamlined objects offer the minimum resistance to the flow of air or liquid around them – they are all curves – and this sleek style improves the aerodynamics of machines such as locomotives. Loewy, however, applied streamlining across the whole of industrial design, restyling familiar objects such as refrigerators to give them a modern look and make them more desirable. One of the most famous objects to get the Loewy treatment was his pencil sharpener, its streamlined metal body finished in chrome. This small gadget had a rounded front tapered to a point like a teardrop: when Loewy dreamed it up in the early 1930s, it looked completely unlike any other piece of office equipment. The shining body recalls the hood ornament from a futuristic automobile, and the metal strut that supports the handle also resembles part of a motor vehicle – perhaps the chrome-plated bracket of a wing mirror. The pencil sharpener symbolized Loewy's progressive ideas about design, and seemed to suggest that design could help the United States move towards a better future in the years following the Great Depression. This may be why Loewy, who was very conscious of the power of publicity and marketing, kept the prototype on display in his office – the device was, in fact, never put into production. With objects like this, Loewy showed Americans that design is not just about function, its symbolic power can also influence people's aspirations. Even something as simple as a pencil sharpener could inspire people to believe in a brave new world.

The diagonal bracket resembles the wing mirror of a car

The meeting point of the body and the rotating end is virtually seamless

The base collects pencil shavings

The stand flares at the base to accommodate the desk attachment

## RAYMOND **LOEWY**

### 1893-1986

Born in Paris, Raymond Loewy moved to New York in 1919, working first as a window dresser at Macy's department store before embarking on a career as a fashion illustrator. His first industrial project was a reworking of a Gestetner duplicating machine in 1929. He used modelling clay to design a sleek body shape, a method he applied to restyling automobiles too, mainly for Studebaker. His practice grew, and by 1951 he had more than 140 staff in offices in the USA, working on everything from logos to aeroplane interiors. His prolific output and design books made Loewy one of the most famous American designers.

# Visual tour

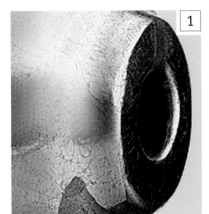

**KEY**

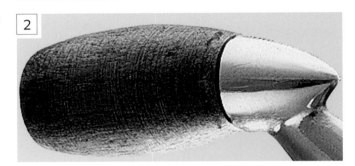

◄ **FRONT** At the rounded front end, the curve tapers at the centre to form a ring around the hole where the pencil is inserted. This combination of curve and circle recalls some of Loewy's designs for streamlined railway locomotives.

▶ **HANDLE** The form of the handle mirrors the pencil sharpener's body: curved at one end and pointed at the other. There is a contrast in shape and texture between the rough material of the handle itself, which is easy for the user to grip, and the metal bracket, with its decorative pointed end.

▶ **REAR** Like Loewy's aerodynamic automobile designs of the 1940s, the pencil sharpener comes to a point at the rear, underscoring the streamlined form characteristic of aerodynamic designs.

The teardrop shape of the body echoes aeroplane design

**IN CONTEXT**

Loewy applied his considerable design skills to a vast range of objects from Lucky Strike cigarette packaging to electrical appliances and furniture. Transportation was a particular interest, and Loewy designed the famous Greyhound Scenicruiser bus and locomotives for the Pennsylvania Railroad, as well as cars. He introduced welded, rather than riveted, body panels for the locomotives, and smoothed out their lines to create more efficient designs. One of the most successful was the S1, which combined a torpedo-like body shape with an artful use of horizontal lines to evoke sleekness and speed. "It flashed by like a steel thunderbolt," Loewy wrote when he saw the train.

▲ **S1 locomotive**, Pennsylvania Railroad

# Anglepoise lamp

1934-36 ▪ LIGHTING ▪ STEEL ▪ UK

SCALE

## GEORGE CARWARDINE

**One of the most ingenious** of all lighting designs, the Anglepoise® lamp was created almost by accident. Its inventor, British engineer George Carwardine, specialized in designing suspension systems for vehicles and became an expert on springs. In the early 1930s, he developed a new form of close-wound spring that could be moved with ease in every direction, but which also remained rigid in any position. When he patented the spring in 1932, he had no idea how to use it. Over the next two years, however, he constructed a lamp mounted on a pair of arms that was flexible enough to be moved backwards and forwards, side to side, or up and down – to focus the light wherever it was needed. Carwardine realized that he had designed the perfect task light for workshops and factories.

The Anglepoise® lamp looked very functional. Mounted on a heavy base, it consisted of a pair of arms linked by an elbow joint and held in position by springs near the bottom. At the top of the upper arm was the lampholder in a simple, rounded metal shade that was just big enough to hold a low-wattage bulb. A 25-watt bulb was sufficient because the light was directed so precisely, which meant the lamp had the added advantage of being economical to run. Carwardine had originally intended the lamp to be used exclusively in an industrial setting, but he soon realized that it would also appeal to domestic users. Working with the lamp's manufacturer, Carwardine produced a revised design with fewer springs, a square, stepped base, and a choice of different colours. The product was launched on the domestic market in 1936 and was an immediate success. Although there have been various modifications, the overall design of the Anglepoise® remains the same. It has inspired many imitations – a sure sign of a classic design.

# Visual tour

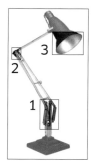

**KEY**

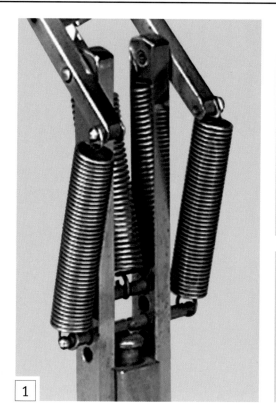

▶ **SPRING ASSEMBLY**
George Carwardine's first lamp had four springs, but in the classic domestic lamp of 1936, a trio of springs holds the arm in place. The loops at the ends of the spring simply hook on to the arm and base in a perfect piece of functional design that also made assembly very straightforward.

1

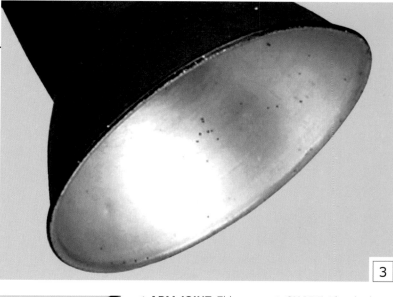

3

◀ **ARM JOINT** This elbow joint enables the upper arm to be hinged up and down, changing its angle relative to the lower arm and holding the arm in position wherever it is needed.

▲ **SHADE** The shade flares out to spread the light so that the focus of the beam is not too narrow. The two-tone finish looks stylish and the pale colour on the inside of the shade is also functional, reflecting the light from the lamp's low-wattage bulb.

2

The lampholder mounting permits a range of movements

The hollow arm carries the electrical flex

The wide shade makes it easy to change the bulb quickly

The heavy, tiered base was inspired by Art Deco design

## GEORGE **CARWARDINE**

### 1887–1948

After leaving school at the age of 14, Carwardine served an apprenticeship in his home town of Bath, UK. Employed in various workshops, he qualified as an engineer by studying in his spare time, and was taken on at the Horstmann Car Company in Bath. By 1916, Carwardine was its Works Manager and Chief Designer. During the 1920s, he had his own business designing and building components for cars, especially suspension systems, but his firm closed in the late 1920s, probably as a result of the Great Depression. Carwardine spent his last years as a freelance consulting engineer and inventor specializing in springs and suspension mechanisms, and although he filed other patents, the Anglepoise® remains his lasting legacy.

> "The Anglepoise® is a minor miracle of balance... Balance is a quality in life that we do not value as we should"

**KENNETH GRANGE**, designer

## ON **DESIGN**

George Carwardine likened the way the Anglepoise® works to the human arm. The springs and the base act like the shoulder muscles, keeping the mechanism steady and allowing it to turn. The joint between the upper and lower arms, with its single direction of movement, functions like the elbow – its effect is to raise or lower the lampholder. The way the lampholder is mounted onto the upper arm, enabling a range of swivelling and tilting movements, is similar to the wrist. Carwardine worked out his design in a series of precise drawings that take into account the various angles involved and the position of the lamp's centre of gravity. The original 1934 model had four springs, but Carwardine found three worked just as well. He took his designs to Herbert Terry and Sons, specialist spring-makers in the British Midlands, and the company soon put the lamps into production.

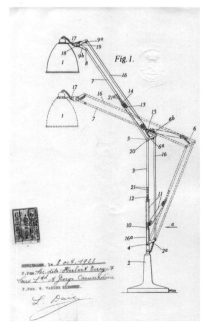

▶ **One of Carwardine's design drawings** for the first version of the Anglepoise® lamp, 1933

# Ekco radio AD65

1934 ▪ PRODUCT DESIGN ▪ BAKELITE ▪ UK

SCALE

## WELLS COATES

**In the 1920s and 1930s, radio ownership** increased dramatically and millions of people came to rely on the wireless, as it was then called, as a source of entertainment, information, and news. Many early radios were in wooden cabinets that blended in with domestic furniture, but in 1934, the British company E.K. Cole ("Ekco") launched a radio with a strikingly different design following a design competition. The winning entry, the AD65, was created by Modernist designer Wells Coates, using Bakelite, the first synthetic plastic and a material that could be moulded with ease to form any shape. Leo Baekeland invented the material in 1907, but when his patent ran out in 1927, Bakelite enjoyed a huge surge in popularity. Companies and designers were free to experiment by changing the formula to make new colours, and made moulds for everything from clocks to combs and jewellery.

The distinctively rounded AD65 radio was one of the most outstanding of all the Bakelite products on offer. Its circular casing made a dramatic statement, but it was also functional – it housed the tuning dial around the edge and the round loudspeaker in the middle, and was economical to manufacture as well, fulfilling Coates' pursuit of "purpose related to purse" in his designs. Coates' AD65, with its streamlined shape, fitted into both Modernist and Art Deco settings and was an international success.

An integral
foot keeps the
radio stable

### WELLS **COATES**

#### 1895-1958

Born in Tokyo, Wells Coates studied engineering in Canada before moving to Britain. He worked as a journalist in London and Paris, where he became interested in the ideas of designers such as Le Corbusier (see p.57). Coates' design career started in in 1928, with a commission to create shop fittings for Cresta Silks. He also designed buildings and wooden furniture for the Isokon company, tubular steel furniture for PEL, and an electric heater, as well as the AD65 radio for Ekco. His work, which combined Modernist forms and up-to-date materials, made him one of the most influential designers of the interwar period.

# Visual tour

The smooth, shiny surface can be wiped clean

The rounded case was produced by casting Bakelite in a mould

**KEY**

◄ **CONTROLS** Three control knobs in the same colour Bakelite as the case give the radio an engineered appearance. Bakelite does not conduct electricity, which made it ideal for switches

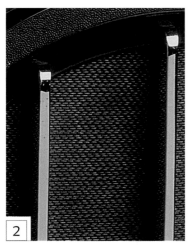

▲ **LOUDSPEAKER GRILLE**
Three vertical stainless steel bars run in front of the cloth-covered loudspeaker at the centre of the case. They give some protection from accidental damage and add a stylish touch to the design

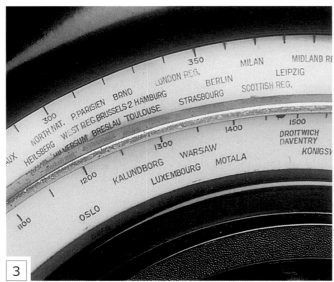

▲ **TUNING DIAL** The moulded case allows plenty of room for a large, semicircular tuning dial, with frequencies and the names of radio stations laid out clearly in legible type. A dial of generous size facilitated precise tuning.

## ON **ARCHITECTURE**

Wells Coates was not a prolific architect, but he designed a small number of highly influential buildings including two Modernist blocks of flats: Embassy Court in Brighton and the Lawn Road flats in Hampstead, North London. The Lawn Road block was commissioned by Jack Pritchard of the Isokon company, and finished in white render with long white access balconies and large windows. Inside were a bar, clubroom, and tiny flats intended as London pieds à terre for professional people. The flats provided an ideal showcase and setting for the Modernist Isokon furniture designed by Coates and his contemporary, Bauhaus designer Marcel Breuer (see p.59) who kept a flat in the block.

▲ **Isokon Flats**, built 1933–34, restored 2004

# Savoy vase

1936 ▪ GLASSWARE ▪ MOULD-BLOWN GLASS ▪ FINLAND

SCALE

## ALVAR AALTO

**During the run-up to the 1937** Paris World's Fair, Finnish glass companies Karhula and Iittala launched a competition for glassware designs to display in the Finnish pavilion. The winner was architect and designer Alvar Aalto, whose entry included a glass vase with a unique and striking form. Unlike most vases, Aalto's design is asymmetrical, its sides forming a series of undulating loops that look more like the folds in a piece of fabric than glass. Its form reminded some people of the rounded shapes of Finnish lakes and others of the designer's name, which in Finnish means "wave". The varying curves of the glass, seen from above or from the side, create gradations in the colour – the vase looks pale and almost transparent where the sides curve gradually, but where they bend to make a tighter curve, the colour is more intense and opaque. When flowers are placed in the vase, their stems fall naturally into the loops, making it easy to create arrangements. The vase was a success at the Paris expo and was adopted by the Savoy, a new luxury restaurant in Helsinki for which Aalto and his wife, Aino, designed the interiors and fittings. First known as the Savoy vase, this beautiful design, still in production by Iittala as the Aalto vase, has enjoyed enduring popularity.

### ALVAR **AALTO**

#### 1898-1976

After studying architecture in Helsinki, Alvar Aalto quickly established a name for himself. At the end of the 1920s, he embraced Modernism with buildings in Finland, such as the Viipuri Library and Paimio Sanitorium. Designed to be informal, light and welcoming, they were hailed as modern masterpieces. Aalto also created furniture, lighting, and fittings for his buildings, and his first commission in Helsinki, the prestigious Savoy restaurant, was a showcase for his furniture as well as the 1937 vase. After World War II, Aalto worked on many projects in Europe and the USA, but his early Finnish buildings, together with his designs for furniture, glassware, and other products, are still his most famous creations.

# Visual tour

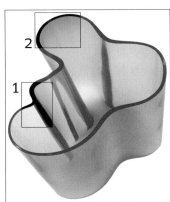

**2**
**1**

**KEY**

**2**

▶ **INTERIOR** The way the glass curves and folds gives the vase a sculptural quality and creates a satisfying pattern of light and dark. Aalto designed the vase to be equally pleasing to look at whether it was empty or full of flowers.

**1**

The mould-blown glass varies in thickness

Green was one of the original colours; the vase was also available in amber, azure, red, smoke, white, and colourless glass versions

◀ **LOOP** There are three large loops around the edge of the vase, each with a distinct shape and each forming a generous, swelling curve. Their fluid shapes, which reflect the designer's love of organic natural forms, give the vase a bowl-like appearance. Aalto is said to have suggested the vase could also be used as a fruit bowl.

## ON **DESIGN**

No one is certain where Aalto got the inspiration for the Savoy vase. He said that one source was the shape of a puddle. Another clue is in the working title the designer gave to his original design – *Eskimoerindens skinnbyxa* (Eskimo woman's leather breeches) – which suggests that Aalto aimed to imitate the complex folds of soft leather. Whatever the actual source of the idea, Aalto clearly wanted to explore the ways in which organic, asymmetrical curves could work when applied to glassware. The curves are clear in the original drawings that he made when working on the competition entry for the Paris World's Fair (right). In these simple sketches, the varying width of Aalto's lines indicates the effect he wished to create with the different thicknesses of the glass when it was blown into a mould.

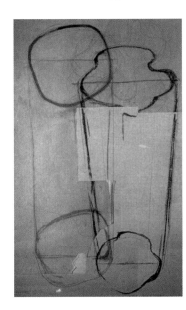

▶ **Sketch for the Savoy Vase,** 1936

## IN **CONTEXT**

Alvar Aalto created many furniture designs that reflected the simplified forms and functional approach of the modern movement. His favoured materials included plywood and birch, which he thought offered more comfort than the tubular steel favoured by other contemporary designers. By bending wood into purposeful curves in chairs such as the cantilevered model Armchair No. 31 (below), he produced furniture that was elegantly modern, yet had a tactile, warm quality. Aalto's inspiration was the Finnish landscape and he had a deep respect for the wood abundantly produced in its forests.

▲ **Armchair No. 31,** 1932

# Kodak Bantam Special

1936 ▪ PRODUCT DESIGN ▪ CAST ALUMINIUM WITH ENAMEL FINISH ▪ USA

SCALE

## WALTER DORWIN TEAGUE

**By the late 1920s, the Kodak brand** was synonymous with photography, and Kodak's films and low-cost cameras sold widely all over the world. In 1928, the company commissioned industrial designer Walter Dorwin Teague to revamp their cameras, in the hope that they would capture an even larger, more fashion-conscious market. Teague began with redesigns of the cheaper Brownie range, but by 1936 he was creating a new, upmarket model with a stylish, Art Deco-inspired striped body that was unlike any previous camera. The compact Bantam Special looked like a fashion item and showed that up-to-date design could be applied to a product that had previously been seen purely in functional terms. The camera was also extremely well made – the aluminium body was very strong, its tapering shape was comfortable in the hand, and the camera was equipped with a high-quality lens and shutter, as well as controls that were precisely engineered. The body opened to reveal the lens,

mounted on a small bellows and surrounded by various controls. A flick of a lever closed the case again, protecting the lens and allowing the user to stow the camera in a pocket or bag. Designed to appeal to photographers who wanted a good-quality, pocket-sized camera, the Bantam Special was, for many, the best-looking camera ever made.

Split-field rangefinder for rapid and easy focusing

The sleek, black enamel finish appealed to stylish customers

A high-quality lens, designed to give good performance at all aperture settings

### WALTER DORWIN **TEAGUE**

**1883–1960**

After studying at New York's Art Students League, Walter Dorwin Teague worked in advertising and publicity before branching out into product design in the mid-1920s. Kodak was one of his first important industrial clients, but he also designed glassware for Steuben Glass and petrol stations for Texaco, plus commissions for major companies such as DuPont, the creators of Nylon, and Ford. By the mid-1940s, he was recognized as one of the most influential designers in the United States, and in 1944 he became the first president of the Society of Industrial Designers.

▲ **Case closed**

An integral plain black cap protects the lens front

# Visual tour

**KEY**

**◄ FILM ADVANCE KNOB** Machined with a rough surface for good grip, the knob advances the paper-backed roll film inside the Bantam Special. The film advance is one of the few controls that is visible when the camera is closed.

**► LENS BARREL** The camera was fitted with a 45mm Kodak lens and a German-made shutter, which was controlled by moving a lever next to the lens barrel. Also engraved on the lens barrel was a depth of field scale to show which parts of a scene would be in focus at a range of different aperture settings.

**► BODY** The tapered aluminium "clam-shell" body was machined to produce an immaculate surface. The gaps between the narrow bands of aluminium were then filled in with black enamel and the resulting striped effect gave the camera a fashionable, streamlined look often found on architecture of the period.

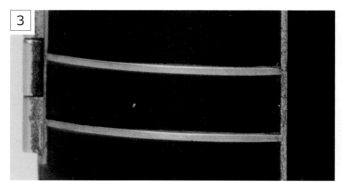

## IN **CONTEXT**

From its origins in the late 19th century and the early slogan "You press the button, we do the rest", the Eastman Kodak Company had always been publicity-conscious and keen to market its cameras as easy to use. By the 1930s, the distinctive red and yellow colours were appearing in publicity and packaging, with the Kodak name in clear, red type so that customers could easily identify the brand. The work of designers such as Walter Dorwin Teague on the camera range made the company aware of the fashion for Art Deco design and prompted them to introduce new typographic styles into their publicity. Tall, modern, italic lettering was integrated with Kodak's standard bold typeface, in advertisements emphasizing the established Kodak virtues of competitive pricing and user-friendliness (right). The combination of well-designed cameras and keen publicity helped Kodak become the biggest name in photography worldwide.

**► Kodak advertisement**, 1935, illustrating one of the company's low-cost Bakelite cameras

# Volkswagen Beetle Model 1300

1938 ▪ CAR DESIGN ▪ VARIOUS MATERIALS ▪ GERMANY

SCALE

## FERDINAND PORSCHE

**Instantly recognizable all over the world**, the Volkswagen Beetle is one of the most successful and enduring of all car designs. It was created as the result of a very specific requirement: Adolf Hitler's demand for a cheap "People's Car" that would carry two adults and three children in comfort, at a speed of 100 kph (62 mph), and at a retail cost of 1,000 Reichsmark (the price of a small motorcycle). The chosen designer was automotive engineer Ferdinand Porsche, who developed a design using a rear-mounted, air-cooled engine in a flowing, aerodynamic body that soon won it the nickname "Beetle". Porsche drew on existing models from Czechoslovakian car manufacturer Tatra for this design. However, the final result – with its spartan interior, simple mechanics, and insect-like body – was distinctive. Its success did not come instantly, though. By 1939, the Volkswagen factory had produced only a small number of prototypes. With the outbreak of World War II, the manufacturer's attention turned to military versions of the car – the jeep-like Kübelwagen (Bucket Car) and the amphibious Schwimmwagen (Swimming Car). After the war, production of the car resumed, this time in volume, and the Type 1 version of the VW Beetle quickly became popular. Over the subsequent decades, the company introduced design improvements – the most obvious were a larger rear window, more powerful engine, and mechanical upgrades such as an improved crankshaft. They did not, however, change the overall concept and shape of the car: it was a winning formula.

◀ **VW logo** The initials for Volkswagen (People's Car) replaced Hitler's preferred name, KdF (Kraft durch Freude) Wagen (Strength-through-Joy Car).

As in traditional automobile styling, the front bumper protrudes from the bodywork

Large wheels improve the ride on rough rural roads

A narrow running board extends beneath the door

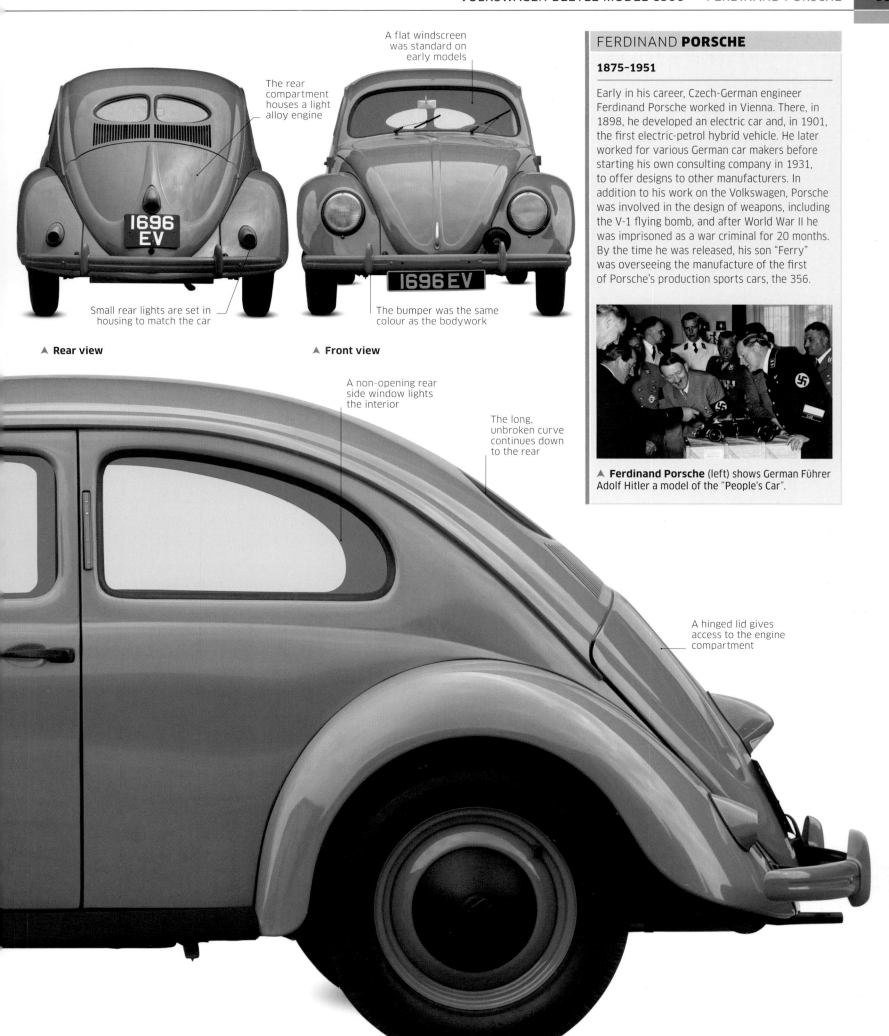

A flat windscreen was standard on early models

The rear compartment houses a light alloy engine

1696 EV

Small rear lights are set in housing to match the car

▲ **Rear view**

The bumper was the same colour as the bodywork

1696 EV

▲ **Front view**

A non-opening rear side window lights the interior

The long, unbroken curve continues down to the rear

A hinged lid gives access to the engine compartment

## FERDINAND **PORSCHE**

### 1875–1951

Early in his career, Czech-German engineer Ferdinand Porsche worked in Vienna. There, in 1898, he developed an electric car and, in 1901, the first electric-petrol hybrid vehicle. He later worked for various German car makers before starting his own consulting company in 1931, to offer designs to other manufacturers. In addition to his work on the Volkswagen, Porsche was involved in the design of weapons, including the V-1 flying bomb, and after World War II he was imprisoned as a war criminal for 20 months. By the time he was released, his son "Ferry" was overseeing the manufacture of the first of Porsche's production sports cars, the 356.

▲ **Ferdinand Porsche** (left) shows German Führer Adolf Hitler a model of the "People's Car".

# Visual tour

**KEY**

▼ **HEADLAMP** The various cars that influenced Porsche's design, such as Tatra and the German-made Standard Superior automobiles, had headlamps mounted on the front fenders or body. These lamps were made as separate units, like the lamps on early cars, and looked as if they had been added onto the body. On the Volkswagen, by contrast, the lamps were built into the coachwork of the front wings, so that they were more integrated into the car's body. The headlamps were protected by the bodywork and the entire front of the car was more aerodynamic.

▼ **REAR WINDOW AND VENTS** The original designs and prototypes of the Volkswagen had no rear window at all and the engine compartment vents continued most of the way down to the bumper. In the revised model that went into production, the designers added a small split window to give the driver some limited rear view. The engine vents fanned out in the space beneath this window and the engine compartment lid. Later versions with bigger, one-piece windows and vents reintroduced into the lid, were more practical but less distinctive.

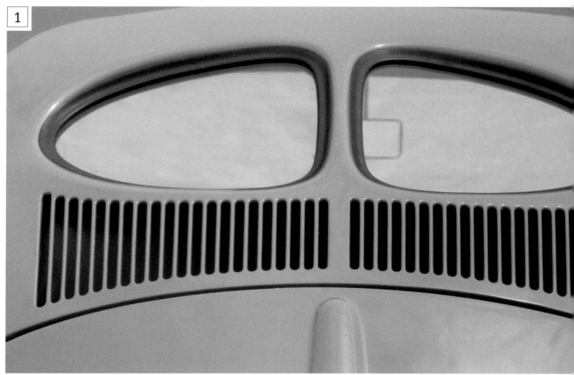

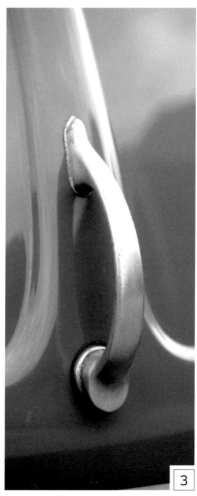

▲ **HORN** When the Volkswagen was designed in the 1930s, it was common for cars to have an air horn consisting of a metal trumpet and rubber bulb mounted on or near the driver's door. Volkswagen equipped their car with an electric horn fitted on the front of the body, and below and to one side of one of the headlamps.

◄ **BONNET DETAIL AND HANDLE** The bonnet contains the luggage compartment in this rear-engined car. It sweeps down to the front bumper, with ribs that run down to the base and meet at the centre. The handle, a plain loop of steel, is mounted at the bottom of the centre rib. The front has a clean, uncluttered look, unlike most contemporary vehicles, which featured a radiator grille, a mascot, and sometimes additional lights.

**5** ◄ **STEERING WHEEL AND DASHBOARD** The dashboard, like the body, is made of metal and the instruments, switches, and warning lights are set into a black panel. The positioning of this panel allows the driver to see the speedometer and odometer clearly through the steering wheel, which has a slender rim and spokes.

## IN **CONTEXT**

During the 1930s, several car manufacturers developed streamlined, rear-engined cars that influenced Ferdinand Porsche when he came to design the Volkswagen. One of the most successful of these manufacturers was the Czechoslovakian company Tatra, whose luxurious T77 had a rear-mounted, air-cooled engine. Tatra also developed a smaller model, the V570, which influenced the shape and layout of the VW, but this was taken only as far as prototype stage. In 1933, a small German manufacturer, Josef Ganz, developed a car called the May Bug, which Hitler is said to have seen at a car show. Like the V570, it combined a beetle-like shape with a rear engine. Porsche improved on both these designs, smoothing out the curves and creating long, flowing lines.

▲ **Tatra V570** prototype

**6** ◄ **INDICATOR** Also called a trafficator, this direction indicator was common on cars in the interwar period. Some vehicles, however, had no built-in indicators, leaving the driver to make hand signals. The Volkswagen's trafficators spring out when they are activated by the driver, and fold away neatly when out of use. The visible edge is the same colour as the car body.

## ON **PRODUCTION**

The Volkswagen was designed to be easy to produce in large numbers. Its air-cooled engine was less complex to put together than a water-cooled type with a radiator, and its rear-engined, rear-wheel-drive design eliminated the propshaft. The car also lacked synchromesh gears and had basic cable brakes. Hitler ordered the construction not just of an enormous, specially designed factory, but of an entire factory town to produce these simple vehicles in volume. Once World War II was over, this facility ceased to make military vehicles and the factory switched over to the production of the "People's Car", turning out vast numbers of cars on state-of-the-art assembly lines.

► **NUMBER PLATE LIGHT** A small light in a metal housing sits above the rear number plate. The shape of this light changed as models evolved over the years, but this original housing was in keeping with the streamlined design of the whole car.

**7**

▲ **By 1946**, the Wolfsburg factory was producing a thousand vehicles a month.

# Knuten candelabra

1938 ▪ METALWARE ▪ SILVER PLATE ▪ SWEDEN

SCALE

## JOSEF FRANK

**Supremely elegant and perfectly balanced**, the Knuten (knot) candelabra is one of the most celebrated pieces produced by the innovative Swedish company, Svenskt Tenn. It was the work of Josef Frank, an Austrian-born designer who emigrated to Sweden in 1933, joining Svenskt Tenn a year later. There, Frank formed a fruitful creative partnership with the owner, Estrid Ericson, and created many successful designs for furniture, textiles, and lighting, which she arranged in attractive room settings in her Stockholm interior design store.

Candelabras for table centrepieces were traditionally constructed with arms radiating out from a central support. Frank adopted a different and very modern approach for his design, dispensing with the base entirely. Three plain, cylindrical candle holders were mounted on the ends of metal tubing that was then bent in a series of graceful curves. Three of these curves dip to form the "feet", which rest on the surface and support the structure. The others intertwine to form the knot that gives the candelabra its name. The fine craftsmanship of the piece was typical of Frank's designs, but the candelabra was also symbolic. The central "knot of friendship" represented the hospitality with which Frank and his wife had been received in Sweden.

### JOSEF **FRANK**

#### 1885–1967

Josef Frank became a successful architect in Vienna, and was a pioneer of modern architecture in the city as well as a designer of furniture, textiles, and carpets. In the 1930s, however, the economic slump, combined with the rise of Nazism, encouraged Frank and his Swedish wife to move to Stockholm. Frank was soon appointed chief designer for the leading Swedish company Svenskt Tenn, a role he held until his death in 1967. His commitment to creating elegant and comfortable interiors played a major part in raising the worldwide profile of Swedish design.

# Visual tour

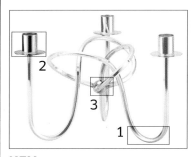

**KEY**

▶ **CURVING METAL SUPPORT**
Although at first glance the series of metallic swirls appear chaotic, the Knuten candelabra is a very stable design, both visually and physically. It rests on three generous curves of metal, ensuring that it stands very firmly on a table or sideboard.

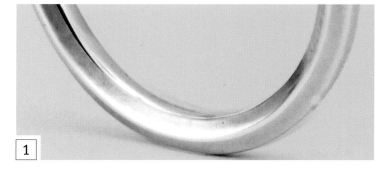

1

▶ **CANDLEHOLDER** This beautifully simple design is made up of a cylinder resting on a metal disc. The crisp, vertical outlines of the candleholder make a striking contrast with the sinuous supports.

2

3

◀ **INTERSECTION** The candelabra is designed to resemble a knot, but the join of the pieces is actually an intersection, where all the supports meet. The join is emphasized and partly concealed by a circular fitting that, in visual terms, acts as a central, pivotal point for all the other elements of the structure.

# "Away with universal styles, away with the equalization of industry and art"

**JOSEF FRANK**

Josef Frank produced a large body of work for Svenskt Tenn, ranging from luxurious sofas and rattan chairs to cupboards and chests decorated with flowers and other motifs. While his pieces were modern in their lines and forms, Frank did not subscribe to the minimalist Modernism of de Stijl (see p.35) or the Bauhaus (see p.46). His work was more colourful and his furniture placed a greater emphasis on both looking and feeling comfortable. Colour played a major part in much of his work, whether in the vibrant textiles that covered his furniture designs or the multicoloured shades that he used on the light fittings of the Floor Lamp No. 2431 (right). In this lyrical, triple-branched design from 1939, there are echoes of the Knuten candelabra.

◄ **Floor Lamp No. 2431**

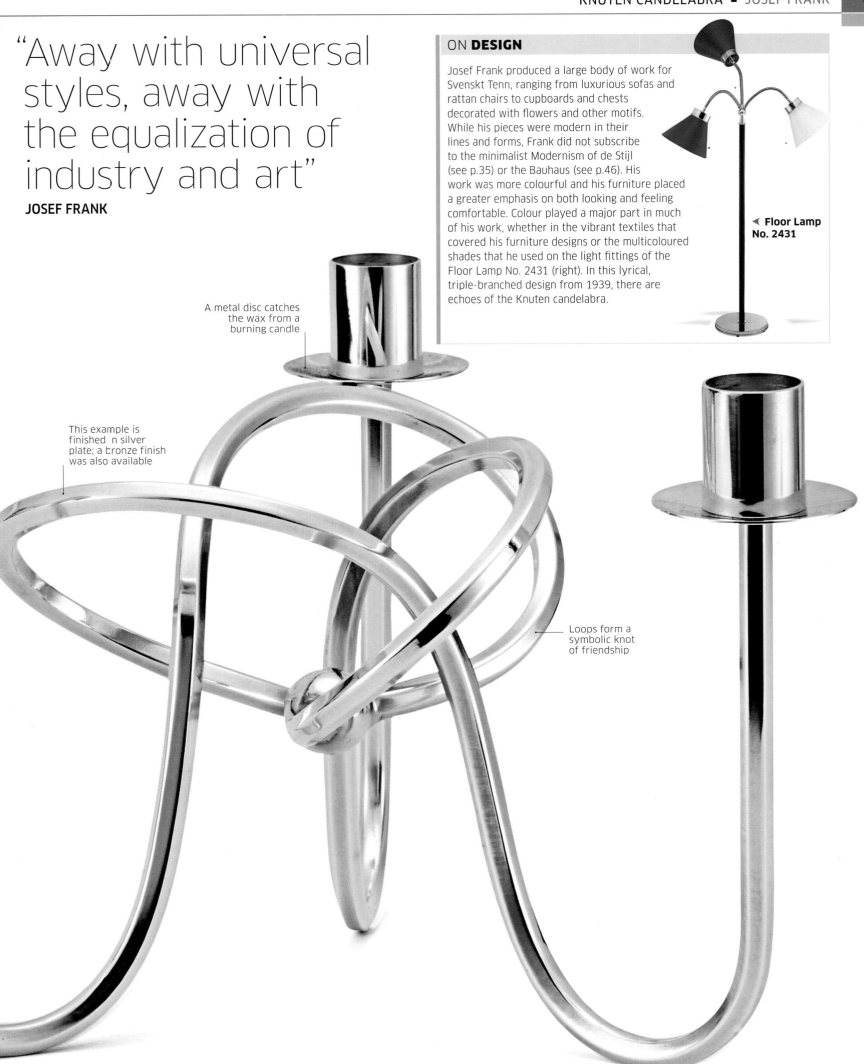

A metal disc catches the wax from a burning candle

This example is finished n silver plate; a bronze finish was also available

Loops form a symbolic knot of friendship

# B.K.F. chair

1938 ▪ FURNITURE ▪ WROUGHT IRON AND LEATHER ▪ ARGENTINA

SCALE

## ANTONIO BONET, JUAN KURCHAN, JORGE FERRARI HARDOY

**A group of designers** working in Argentina in the 1930s created one of the most enduring chairs of the 20th century. The B.K.F. chair (the name is made up of the first letters of the designers' surnames) shows that good design need be neither expensive nor luxurious. The chair frame consists of just two lengths of iron rod. Each rod is artfully bent at a precise angle and the two pieces are joined together so that four of the bends stand on the floor to stabilize the chair. The other four support the cover at each corner. The result is a light and comfortable chair.

The original covers for the chair were made of leather and were produced by a firm of saddlers, whose regular work – making saddles for polo ponies – guaranteed a strong, high-quality finish. At first, the chair was scarcely known outside Buenos Aires but in 1943, it was included in a major exhibition organized by the Argentine Ministry of Culture, where it was awarded first prize and attracted the attention of Edgar Kaufmann Jr., who was curator at The Museum of Modern Art in New York. Kaufmann bought two of the chairs – one for the museum's collection and another for his father's house, Fallingwater, which had been designed by Frank Lloyd Wright and was one of the most famous modern houses in the world. The B.K.F. chair was now firmly in the public eye and it became popular all over the world. It was the chair of choice for many design-conscious consumers furnishing their homes on a budget and it spawned many cheaper imitations, often with fabric rather than leather covers.

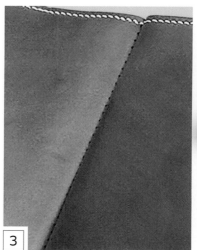

## ANTONIO **BONET**

### 1913-89

The Catalan architect, urban planner, and designer Antonio Bonet was born in Spain. He studied architecture in Barcelona, then worked in the offices of the Catalan architect and planner Josep Lluís Sert, and for Le Corbusier (see pp.56–57) in Paris.

In 1938, Bonet moved to Buenos Aires, Argentina, where he founded the progressive Grupo Austral with other architects, including Juan Kurchan and Jorge Ferrari Hardoy. He worked on various city-planning projects (including the South Buenos Aires Urban Development Plan), as well as designing many houses in the Modernist style, such as the Casa Berlingieri (1946) at Punta Ballena, Uruguay. He also designed furniture, and taught architecture as a visiting professor at Tucumán National University in Argentina. In 1963, Bonet returned to Spain and continued to practise as an architect in Girona and Barcelona. His best-known building from this period is the prominent Urquinaona Tower (1971) in Barcelona.

# Visual tour

**KEY**

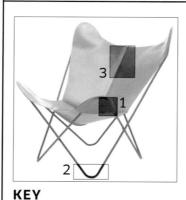

**◀ COVER POCKET** Each corner of the cover has a sewn pocket into which a curving section of the frame fits. These four pockets hold the cover in place, so that it is suspended on the frame like a hammock. The cover can be slipped off when the chair has to be cleaned or stored.

1

2

**▲ METAL FEET** The metal rods that make up the framework of the chair are bent to the correct angle for optimum stability. The slightly flattened section of steel at the base acts as a foot that rests firmly on the floor.

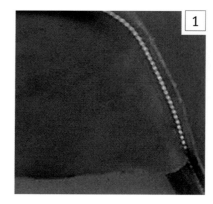

3

**▶ BACK SEAM** The cover is made of several pieces of leather that have been stitched together. This long, vertical seam creates slight angles in the two halves of the cover, so that it can follow the contours of the user's back.

The robust leather cover has saddle-stitched seams

Iron rods give the chair a light, yet strong frame

The rods are welded together at the points where they cross

## IN **CONTEXT**

When the B.K.F. chair was first launched, many people were surprised by its radical appearance, but its design was not entirely new. The manufacturers of campaign furniture used by armies on the move had developed a similar type of chair in the 19th century. One example was the Tripolina, a seat used by Italian officers in North Africa. It had a folding wooden frame that was angled in a similar way to the B.K.F. chair, and a fabric seat. Another design that resembled the B.K.F., which was made in Britain, became popular as a folding chair for travellers, as well as for use in the home. No one knows whether Bonet and his colleagues knew about these chairs, but some historians think it likely.

▲ **Folding campaign-style chair**, in use in around 1925.

■ **Emerson Patriot radio** Norman Bel Geddes

■ **Harper's Bazaar magazine** Alexey Brodovitch

■ **Tupperware** Earl Tupper

■ **Vespa** Corradino D'Ascanio

■ **Penguin paperback covers** Jan Tschichold, Edward Young

■ **Coffee table** Isamu Noguchi

■ **Fazzoletto vase** Paolo Venini, Fulvio Bianconi

■ **Atomic wall clock** George Nelson

■ **Calyx furnishing fabric** Lucienne Day

■ **Festival of Britain symbol** Abram Games

■ **Birch platter** Tapio Wirkkala

■ **Kilta tableware** Kaj Franck

■ **Diamond armchair** Harry Bertoia

■ **Pride cutlery** David Mellor

■ **Fender Stratocaster** Leo Fender

■ **M3 Rangefinder camera** Leica Camera AG

■ **Butterfly stool** Sori Yanagi

1940–1954

# Emerson Patriot radio

1940 ▪ PRODUCT DESIGN ▪ CATALIN (PLASTIC) ▪ USA

SCALE

## NORMAN BEL GEDDES

**In the 1930s, chemists discovered** ways of adding dyes to the resins used in making plastics, and a whole world of colourful materials was opened up to manufacturers. At the same time, designers in the United States were developing new approaches to styling – covering everyday items, such as refrigerators, in striking, often streamlined, cases that made them seem new and exciting. The use of coloured plastic and the restyling process came together in the Patriot radio, designed by Norman Bel Geddes for the American Emerson company and released in 1940.

Bel Geddes, who was one of the best-known exponents of aerodynamics and streamlining, based his design around the colours of the American flag, with a striped loudspeaker grille and control knobs adorned with stars. The material was Catalin, a plastic that was cast in a lead mould, baked for a period of several days, and then polished to a rich, lustrous finish. The result, a world away from the typical brown Bakelite radios of the period, set a new standard for electrical appliance design. It also appealed to the American psyche. In 1940,

when the Patriot was put on sale, World War II was raging in Europe and people wondered whether and when the USA would join the conflict. This symbol of American identity and technology, launched with a major advertising campaign that also celebrated Emerson's 25th anniversary, played its part in helping to maintain national optimism.

The polished surface makes the colour glow

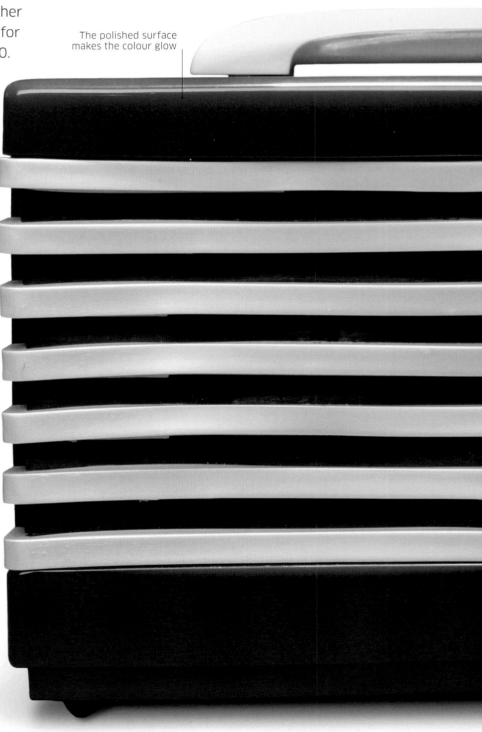

### NORMAN **BEL GEDDES**

#### 1893–1958

After ten successful years as a theatrical designer, Norman Bel Geddes opened an industrial design studio in 1927, applying his adventurous approach to a wide range of products from desks to cocktail shakers. At the same time, he wrote articles and a book, *Horizons* (1932), which did much to popularize streamlining as both a practical and a functional approach to design. Bel Geddes also designed a city of the future for the General Motors pavilion at the 1939 New York World's Fair, and a number of unbuilt but influential projects, such as a teardrop-shaped car and an enormous airliner. In another book, *Magic Motorways* (1940), Bel Geddes proposed a forward-thinking road system for the USA that seemed to anticipate 1960s' freeways.

## ON **DESIGN**

Norman Bel Geddes was a fine draughtsman, carefully working out many of his designs in deta led pencil drawings. When colour was important to the project, he made sketches with coloured crayons – a technique that helped him plan the lighting of his Futurama exhibit for General Motors at the 1939 New York World's Fair. This crayon sketch, which shows an early design for the Patriot

radio, is not a highly finished drawing, yet it gives a clear idea of how the three colours would be used. The small red stars just seen at the top of the tuning dial did not, however, make it into the final design.

▶ **Crayon sketch** of the Patriot radio, 1939

# Visual tour

**KEY**

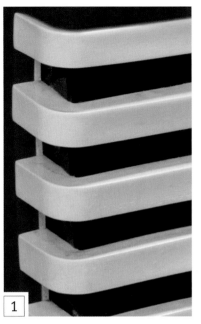

▶ **STRIPED GRILLE** The corner of the striped grille shows how Bel Geddes used the casting process to make the corners of the white stripes curve. This is a perfect piece of streamlining in miniature, similar to the large-scale effects that architects were achieving on the corners of buildings.

◀ **CONTROL KNOB** Five-pointed stars adorn the tuning and volume controls on the front of the radio. The design sketch shows that Bel Geddes considered defining the stars in white against a red background, but he eventually settled on using the casting technique to make the stars stand out in shallow relief.

Bel Geddes provided a carrying handle, even though the radio was mains-powered

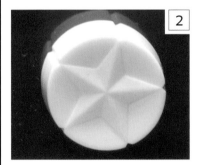

▲ **TUNING DIAL** When laying out the tuning dial, Bel Geddes made good use of the three colours. The numbers stand out in blue sans serif type against the white background of the dial and the red pointer is easy to read against the blue background.

◀ **EMERSON LOGO** The radio bears the Emerson logo. This shows a treble clef with the centre part adapted to depict a stylized radio microphone of the type mounted on springs to isolate it from vibrations. With this small device, technology and music are cleverly brought together.

# Visual tour

**KEY**

**▶ COLOUR** The model's lips provide the only colour on the cover apart from the masthead and other lettering. They are printed as blocks of primary yellow, red, and blue, as well as a dark green that matches the green of the words "Harper's" in the top left corner and "College fashions" in the bottom right. The stylized, sensuous shapes suggest butterflies or flowers.

**▼ SCRIPT TYPEFACE** *Harper's Bazaar*, founded in 1867, already had a long history by the time Brodovitch started work there. He retained the traditional script typeface as a symbol of the magazine's illustrious past, but kept the words set in the font relatively small, so that the modernity of the cover was able to shine through.

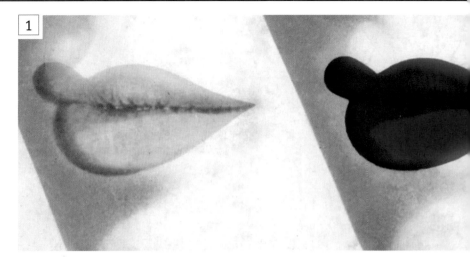

2

1

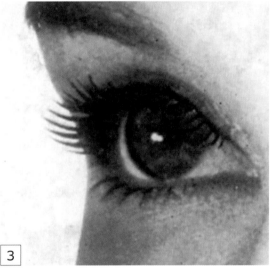

3

**◀ ABSTRACTION** Most magazine covers of the period featured a head-to-toe or portrait photograph of a model, but Brodovitch chose to concentrate on elements of the face for this cover. By repeating the model's three-quarter profile eight times in an overlapping pattern, he created a semi-abstract, tonal image. Half the face is in shadow and all attention is focused on the eyes and the brightly coloured lips.

## IN CONTEXT

Alexey Brodovitch commissioned many leading photographers to work for *Harper's Bazaar*. Some of them, including Irving Penn and Lisette Model, had attended Brodovitch's Design Laboratory sessions. Another pupil, Richard Avedon, began working for Brodovitch in his early twenties and became well known for his shots of smiling, animated models and his sensuous depiction of fabrics. His work came to define what was stylish in American fashion photography for around 50 years.

Some of Avedon's best work for *Harper's Bazaar* showed a figure creating an unusual shape cut out against a white background (right). These bold shapes, created by a dramatic hat or a swathe of brightly coloured fabric, owed much to the influence of Brodovitch and his love of Surrealist art. Avedon worked for *Vogue* as well as for *Harper's Bazaar*, but his unique vision was not confined to fashion – he was also a renowned portrait photographer and chronicler of activists in the USA.

Brodovitch and his successors, among them Herbert Wolf, also played around with the typography used on the cover. Brodovitch would often change the word that was set in large type and occasionally switch typeface, although adhering to the same general style.

▲ *Harper's Bazaar* **cover**, 1956, featuring Audrey Hepburn. Photograph by Richard Avedon

▲ *Harper's Bazaar* **cover**, 1959. Photograph by Richard Avedon, art director Herbert Wolf

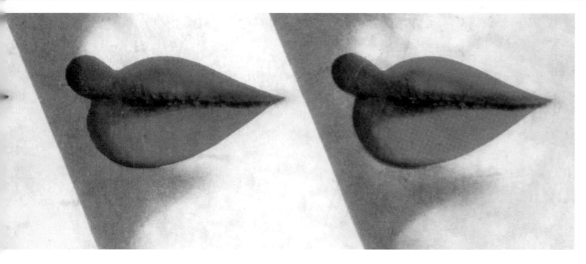

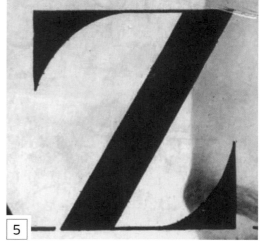

5

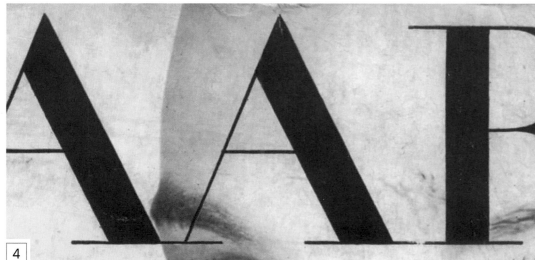

4

▲ **TYPOGRAPHY** Brodovitch favoured traditional typefaces of the kind known as Didone, in which there is a marked contrast between the thick and the thin strokes. The thick strokes are mostly vertical and the thin, hairline strokes generally horizontal. These features are exaggerated in the bold capitals used for the word "Bazaar", emphasizing the magazine's name and creating a strong identity.

◀ **LETTER SPACING** The designer increased the impact of the type by setting the letters so close together that the serifs (the small end strokes or "tails") actually join at some points along the bottom. This creates the effect of underlining the magazine's title in places and avoids the "airy" quality the letters might have had if they had been set farther apart.

## ON **DESIGN**

Under Alexey Brodovitch, the inside pages of *Harper's Bazaar* were just as striking as the covers. Brodovitch was unusual in thinking of layouts in terms of double-page spreads rather than single pages. This meant that he was very conscious of the way a pair of pages worked together, mirroring or forming a contrast to each other, and he often let a full-page image extend across the central gutter between the pages and encroach upon the facing page. Some of his most distinctive layouts, however, relied on pairing a strong image on one page with an innovative arrangement of type on the facing page, as on the right, where the text is set diagonally to imitate the line of the model's dress. Brodovitch completed the effect by using a heading in italics, so that all the type on the right-hand page is leaning at the same angle, creating a layout that is both surprising and pleasing.

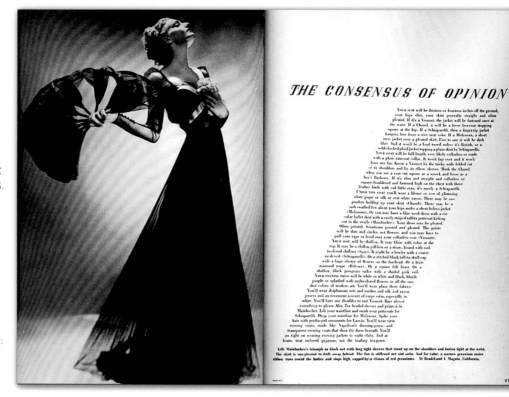

▶ *Harper's Bazaar* **double-page spread**, 1936.
Photograph by Man Ray

# Tupperware

1946 ▪ PRODUCT DESIGN ▪ PLASTIC (POLYETHYLENE) ▪ USA

## EARL TUPPER

➤ **Flask**

The range was available in pastel colours and white

➤ **Creamer jug**

➤ **Storage container**

**Tupperware containers and their many imitators** are so widely used for keeping food fresh that we tend to take them for granted. Yet the inventive use of flexible plastic – still a new material in the 1940s – combined with the ingenious, airtight seal earns them a place among the great designs of the 20th century.

Tupperware had a long gestation. Its story began when American inventor Earl Tupper founded a company in the late 1930s and began to experiment with new forms of plastic. He developed a way of processing polyethylene slag (a waste product of oil refining that was hard, black, oily, and smelly) into a material that was durable, hygienic, and translucent. Unlike most early plastics, which were brittle, the new material was pliable and could be moulded into shape. Looking at the way that metal lids fitted onto paint cans, Tupper developed a similar type of seal that he could mould in the plastic. This meant that his containers could be both airtight and – because of the plastic's flexibility – easy to open and close.

By combining his two inventions – polyethylene plastic and the airtight seal – Tupper created a winning formula, and his company began to produce lightweight, nonbreakable containers in attractive pale colours. Tupper thought he had found the solution to a whole range of kitchen-storage problems, but the containers did not sell well in hardware and department stores. It was only when the company began to market the range exclusively through "Tupperware parties" in people's homes that the products became successful. From then on, the ingenious, pastel-coloured, moulded containers, which included basins, cups, jugs, and boxes, many with airtight lids, became hugely popular in the postwar United States and elsewhere.

### EARL **TUPPER**

#### 1907-83

Earl Tupper was born in the US state of New Hampshire. His parents had a small farm and he began his working life there before setting up a landscaping and tree surgery business in the early 1930s. In 1937, he took a job with the chemical company DuPont, where he learned about plastics, and in the following year set up on his own. Initially producing items for the defence industry, he began to develop domestic products, such as a nonbreakable plastic bathroom tumbler. Thanks to the invention of Tupperware, his company eventually became so successful that Tupper sold it in 1958 for US$16 million and retired to an island off Mexico.

# Visual tour

**KEY**

> **SEAL** A groove in the lid forms an airtight seal with the walls of the container. The container can be sealed by simply pressing gently on the lid to remove excess air, and opened just as easily using the protruding rim on the lid.

> **JUG LID** The polyethylene plastic used for Tupperware is ideal for simple shapes, such as this jug lid. The raised handle, with its smooth, rounded edges, is comfortable to grip and easy to pick up.

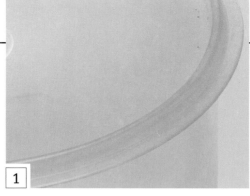

**1**

**2**

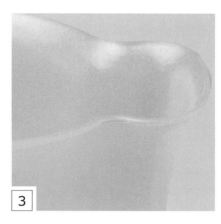

**3**

▲ **CUP HANDLE** The plastic used to make the cup has simply been extended to create the cup handle. By making the cup in this way, there was no need to make a complicated, looped handle. It also adds to the capacity of the cup itself, as the handle can hold liquid too.

## "What's left... stays right"
**TUPPERWARE ADVERTISEMENT**

◄▼ **Stacking cups**

The translucent plastic made it possible to see whether a container was full or empty

The flat lids made it easy to stack containers on top of each other

## IN **CONTEXT**

In the late 1940s, a few people, including saleswoman Brownie Wise, began to include Tupperware in a range of homeware that they were selling through parties in their houses. When Tupper heard of the idea, he hired Wise and Tupperware parties began. This method of selling appealed to many women, as it gave them the chance both to work from home and to socialize. The products proved popular and the parties were hugely successful for Tupperware.

▲ **1960s advertisement** for Tupperware parties

# Vespa

1946 ■ VEHICLE DESIGN ■ HIGH-TENSILE STEEL FRAME ■ ITALY

SCALE

## CORRADINO D'ASCANIO

**AFTER WORLD WAR II**, the Italian economy was in disarray. Piaggio's factory, like those of many other manufacturers that had produced aircraft during the war, had been destroyed by bombing. Looking for ways to rebuild his business, Enrico Piaggio realized that there was a market for personal transport. The poor state of Italian roads and the restricted budget of most consumers, however, made car-manufacturing an unattractive option, so Piaggio decided to develop a motor scooter to provide Italians with a cheap means of transport. Unimpressed with his company's first efforts, he hired an aeronautical engineer, Corradino D'Ascanio, who was already working on a scooter design. D'Ascanio, who disliked motorcycles, because of their oily engines and frame that the rider had to straddle, came up with a new design based on the American Cushman scooters that had been used by troops during the war

Although the small wheels, step-through frame, and rear engine were influenced by the Cushman, the elegant, streamlined shape of D'Ascanio's design reflected the rounded, aerodynamic contours of aeroplanes. The elimination of the oily

drive chain found on motorcycles, and the wraparound bodywork that covered the engine and kept out the wind, proved very appealing to smart young Italians. Women, in particular, appreciated the step-through design, as it made the vehicle easy to ride when wearing a skirt. The scooter's narrow-waisted appearance and the high-pitched buzz of its engine reminded Piaggio of a wasp (*vespa* in Italian), so by the time it was launched at the 1946 Milan fair, the name had stuck. By bringing cheap, stylish transport to millions of people across Europe and beyond, the Vespa became a design classic.

Fuel cap

Both the form and steel cladding of the body are reminiscent of aircraft design

## "Sembra una vespa!"

(It looks like a wasp!)

**ENRICO PIAGGIO**

Headlamp

Side axle

▲ **Front view**

## CORRADINO **D'ASCANIO**

### 1891–1981

Corradino D'Ascanio graduated in engineering and joined the Italian army, where he worked on aeroplane engines and fitted radio equipment to aircraft. After World War I, he developed helicopters, work he continued with Piaggio during World War II. In 1945, he was approached by Italian manufacturer Fernando Innocenti to design a motor scooter, but Innocenti and D'Ascanio clashed, and D'Ascanio took his ideas to Piaggio. After the Vespa project, D'Ascanio continued to develop helicopter designs, wrote scientific publications, and taught at the University of Pisa.

➤ **Name badge** Founded in the 19th century as a manufacturer of railway rolling stock, Piaggio had made aircraft since World War I. When buyers saw the familiar badge on the Vespa, they were reassured by the company's long engineering heritage.

The combination of a high seat and small wheels make the Vespa comfortable to ride

The design of the headlamp housing shows the influence of streamlining

Parallel ribs provide grip for the rider's feet

The front mudguard covers almost half of the wheel

**P**
**PIAGGIO**
**GENOVA**

# Visual tour

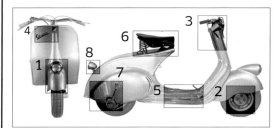

**KEY**

1

▶ **WHEEL AXLE AND MUDGUARD** The front wheel of the scooter is shielded by a beautifully streamlined mudguard with an integrated headlamp in a teardrop-shaped mounting. Beneath the mudguard on one side of the wheel, the axle and part of the suspension mechanism are just visible. Revealing the aeronautical background of D'Ascanio and Piaggio, this side axle is designed in a similar way to the undercarriage of an aeroplane.

2

▲ **FRONT WHEEL MOUNTING** The Vespa has a side axle rather than the front fork common on motorcycles. This makes it easy to remove the wheel from the opposite side to the axle, so maintenance and changing tyres are very straightforward. To free the tiny 20-cm (8-inch) wheel, the four wheel nuts simply have to be undone – a task most Vespa owners could carry out without calling out a mechanic.

3

4

▲ **FRONT FARING AND BADGE** The rising front of the Vespa was one of its most distinctive design features. It became a major selling point, especially for city workers who wanted to protect their clothes from airborne dust and dirt. The scooter's name, underlined in a curving, cursive script, echoes the overall streamlined styling of the Vespa

◀ **HANDLEBAR AND CONTROLS** The main controls are placed on the handlebar, within easy reach of the rider's fingers. One lever activates the front brake and another works the clutch, which is used in conjunction with a twisting grip to change gear. The throttle is also on the handlebar.

► **FLOOR** The metal floor area, ridged for maximum grip, also forms the structural bridge between the front and rear of the scooter. Underneath, additional metal strips reinforce the thin steel, so that the scooter does not buckle when being used.

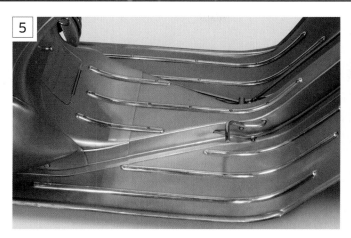

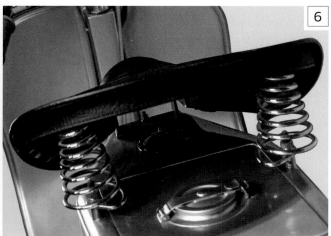

◄ **SEAT** The seat is large and well sprung, to help cushion the rider on Italy's rough, postwar roads. A hinge allows the rider to lift up the seat to remove the fuel cap and fill up. The tank is directly above the engine, which makes it easy to fill.

▲ **REAR BODY AND TAIL LIGHT** The Vespa's body sweeps down from the seat to the rear end in one streamlined curve, a shape that was familiar in car design but unique in scooters at the time. In the middle of this expanse of gleaming metal-work, the generously proportioned circular tail light is clearly visible on top of a sturdy metal bracket.

## IN **CONTEXT**

Although only about 2,500 Vespas were sold in the first year of production, Piaggio were selling some 60,000 per year by 1950. Initially, most sales were in Italy, but the Vespa was given a major boost by the 1953 movie *Roman Holiday*, whose fashionable stars, Audrey Hepburn and Gregory Peck, rode tandem on a Vespa through the streets of Rome. In the 1950s and 1960s, Vespas became popular with everyone, from film stars on studio lots to teenagers who wanted cheap transport. Clean, inexpensive, and easy to ride and maintain, the Vespa was adored for its perfect combination of economy and style.

▲ **A phenomenal success with both women and men,** around 2,500 Vespas were sold in 1947, more than 10,000 in 1948, 20,000 in 1949, and around 60,000 in 1950.

▲ **ENGINE HOUSING AND STARTER** The first Vespas had a 98cc engine that, although small, was more powerful than those of its competitors. The rider started the scooter by pressing the starter lever down with his foot. The distinctive vents in the side of the housing let in air to cool the engine. Piaggio soon added fins to the flywheel to fan the engine and increase the cooling effect.

# Penguin paperback covers

1947-49 ■ GRAPHICS ■ PRINT ON PAPER ■ UK

SCALE

## JAN TSCHICHOLD, EDWARD YOUNG

**Penguin paperbacks were launched** in 1935 by British publisher Allen Lane, with the aim of producing interesting editorial content at an affordable price. The company's first production manager, Edward Young, came up with the original idea for the series cover layout, which was simple and distinctive. The design used three horizontal bands, with the title in the central white band and the publisher's name and symbol in the upper and lower coloured sections. Most of the type was in the 1920s Gill Sans typeface, designed by Eric Gill.

Penguin used this layout for ten years, but not in a systematic fashion, and some cover designs varied from this style. In 1947, German typographer Jan Tschichold was hired to improve and standardize the design of the books. Instead of embarking on a radical redesign of the covers, Tschichold made a number of fine, but perfectly judged, adjustments. All the type on the front covers was now set in Gill Sans, with consistent spacing between the letters; the title was the only element in bold type. Tschichold drew a new Penguin symbol and added a short rule between the title and the author's name. He also developed typographic styles for long titles, and for descriptive copy and subtitles. These changes were subtle, but they had a profound effect. The composition rules Tschichold established were used across the board, creating a clear identity that contributed significantly to Penguin's success as a publisher and resulted in wide recognition of the brand.

---

### JAN **TSCHICHOLD**

#### 1902-74

The son of a German signwriter, Jan Tschichold was influenced by the Bauhaus (see p.46) and Russian graphics. He was a leading Modernist typographer by the late-1920s, advocating the use of sans serif type and non-centred layouts in his influential book *Die Neue Typographie* (The New Typography), 1928. When the Nazis came to power in Germany, he moved to Switzerland and gradually changed his outlook, embracing more traditional typography and graphics. As well as his work at Penguin from 1947-49, Tschichold designed several typefaces (including the popular Sabon) and books, taught, and wrote extensively on calligraphy, typography, printing, book design, and related topics.

## Visual tour

**KEY**

▶ **PENGUIN** The company's "dignified but flippant" symbol had taken several forms (leaning to one side or extending its flippers). When Tschichold modified the design, he gave the bird a smooth outline, sleek flippers, and a head that could be turned to the left or right. This penguin remained in use for several decades.

◀ **TYPE** Penguin had originally used a serif face, Bodoni Ultra Bold, for the type inside the distinctive quartic shape. Tschichold chose a sans serif font and set the publisher's name in the same Gill Sans capitals used on the rest of the cover. The spaces between letters and words were precisely calculated.

▲ **COLOURED BAND** The broad horizontal bands were the elements that first made Penguin paperbacks instantly identifiable, even when seen from a distance. Wisely, Tschichold retained them, adding a subtle line along the edges (although this detail was later abandoned).

For extra clarity, some titles had this identifying line in coloured type

The weight of the quartic's outline varied, like a calligraphic line

PENGUIN BOOKS

FICTION

DON'T, MR DISRAELI

CARYL BRAHMS

S.J.SIMON

FICTION

COMPLETE    UNABRIDGED

1/6

This line reassured readers that the cheaper paperback edition was not a shortened version of the original

The paperbacks originally had dust jackets, which showed the price

## ON **DESIGN**

The different-coloured jacket bands of Penguin's books indicated the various publishing categories. Fiction titles in the company's familiar orange predominated, and these were joined by crime and mystery novels in green, biographies in dark blue, essays in purple, and travel books in cerise, amongst others. The company's large non-fiction list was published under the Pelican imprint, with covers banded in bright pale blue. This colour-coding system (especially for fiction, crime, and Pelicans) endured for decades.

▶ **Four colour-coded covers by Tschichold**

## IN **CONTEXT**

Since the 1940s, Penguin covers have had a long and steady evolution. A major departure in the 1950s was the use of vertical, rather than horizontal, bands of colour and a large central area for an illustration. Tschichold initiated the design, which was completed by typographer Hans Schmoller in the early 1950s. In 1962, Penguin's new art director, Germano Facetti, commissioned a different grid (initially for the crime series) from Polish-born designer Romek Marber. His grid arranged the text at the top of the cover between horizontal rules, leaving plenty of space below for the illustration. From this point, strong, graphic artwork commissioned or produced by Facetti became a major part of the covers, once more making Penguin's titles stand out in bookshops.

▶ **The New Men**
The vertical grid developed by Hans Schmoller and used in the 1950s and 60s was designed to take a small illustration, usually in black and white. Later, the covers were made more pictorial, and the artwork extended into the coloured areas on either side of the central panel.

▶ **Dreadful Summit**
Romek Marber's grid, the distinctive green background, and a graphic two-colour image draw attention to this 1964 mystery novel. It was one of dozens of books in the Penguin Crime series that benefited from the design skills of Marber and Facetti.

# Coffee table

1947 ▪ FURNITURE ▪ WOOD AND GLASS ▪ USA

SCALE

## ISAMU NOGUCHI

**When Japanese-American artist** Isamu Noguchi created this beautiful low table, he brought together sculpture and furniture in a new and striking manner. Structurally, the table appears very simple, but the way that its identical wooden supports perfectly echo and balance each other reveals Noguchi's mastery of sculpture. These interlocking supports reflect images from Surrealist art, and their bold, biomorphic shapes catch and hold the eye, especially as they are visible through the top of the table as well as from all sides. The top, although made of 19mm (three-quarter-inch) thick glass to stabilize the table, seems to float. Broadly triangular, its rounded corners and its one gently curving side give the glass an organic quality that harmonizes with the flowing lines of the base.

The genesis of the table was complex. In 1939, Noguchi produced a one-off table along similar lines for A. Conger Goodyear, the President of The Museum of Modern Art in New York. Soon afterwards, British-American architect and interior designer T.H. Robsjohn-Gibbings asked the sculptor to design a table for him and Noguchi sent him a plastic model based on the one he had created for Conger Goodyear. Noguchi then heard that Robsjohn-Gibbings had begun marketing a variant of his design, and when Noguchi objected, Robsjohn-Gibbings replied that "anyone could make a three-legged table". Stung, Nioguchi made his own

variation on the design and this version was used by designer George Nelson (see p.108) to illustrate an article called "How to Make a Table". The Herman Miller furniture company took on the design, and its surrealistic Modernism soon captured people's imagination. The coffee table is still manufactured – a tribute to Noguchi's remarkable design skills – and is available in a variety of woods, from a pale maple to a striking black-stained ash.

The top is made of heavy plate glass

### ISAMU **NOGUCHI**

#### 1904–88

Sculptor and designer Isamu Noguchi, was born in Los Angeles to a Japanese father and an American mother. He studied cabinet-making in Japan, then became assistant to the sculptor Constantin Brancusi in Paris, before returning to the USA to pursue his own work. Noguchi applied his remarkable skills to a wide range of products, such as the Nurse Bakelite intercom (1937) for Zenith, furniture for Herman Miller and Knoll, and distinctive lighting designs. He also created stage sets for dance companies and designed parks and gardens. His sculptural work ranged in style from realistic portrait busts to Surrealist-inspired pieces and large-scale abstract works.

# Visual tour

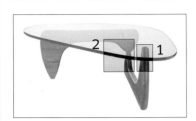

**KEY**

1

▲ **SUPPORT** One corner of the table top is held in place solely by the small, flattened end of the upward-pointing support. This increases the sense of transparency, making it easier to appreciate the structure below.

2

◄ **MEETING POINT** The two pieces of shaped wood that support the table top meet at one end, joined by a cleverly concealed steel rod. The two supports are set at an angle to each other, so the effect is one of slight asymmetry, in keeping with the organic nature of the design.

The legs pivot on a concealed metal rod

The curved edges harmonize with the shapes of the base

The heavy wooden supports ensure stability

## ON DESIGN

The periods Noguchi spent in Japan both as a child and an adult inspired a love of Japanese arts and crafts. He designed ceramics that were produced in Japan and worked on paper lamps, based on traditional Japanese fishing lanterns, which he called Akari. In Noguchi's hands, these lamps became delicate sculptural forms. He stated, "Everything is sculpture. Any material, any idea without hindrance born into space, I consider sculpture." His Akari lamps are still produced in the same way – from translucent paper (originally Japanese mulberry-bark paper) stretched over bamboo ribs – and by the original company, which is based in Gifu, Japan. The framework of the lamps is made from metal rods and wire.

▲ **Akari lantern**, Isamu Noguchi, 1951

## IN CONTEXT

Throughout his long and productive career, Noguchi created sculpture in a variety of styles and materials. Under the influence of the sculptor Brancusi, pioneer of simplified forms, he began to sculpt in stone, but soon progressed to working with metals, wood, fibreglass, and even plastic. Much of his work from the 1930s and 1940s shows the same curvaceous shapes (abstract, but sometimes reminiscent of human or animal forms) as the coffee table. By the 1950s, he was producing more rectilinear pieces, which were often monumental in size. His later work also includes rock-like objects suited to the gardens he designed for corporations and public spaces. For all its variety, Noguchi's work displays a profound respect for materials, a deep commitment to craftsmanship, and a love of simple, expressive forms.

► **Orpheus**, a sculpture in cut and bent aluminium, Isamu Noguchi, 1958

# Fazzoletto vase

1948-49 ▪ GLASSWARE ▪ BLOWN, CASED, AND FILIGREE GLASS ▪ ITALY

SCALE

## PAOLO VENINI, FULVIO BIANCONI

**The island of Murano, off the coast of Venice**, is one of the most famous centres of glass-making in the world, with a history that dates back to the 13th century. In 1947, the Italian designer Fulvio Bianconi, who had been commissioned to create a series of perfume bottles, visited the island and became fascinated with the glass-making process. The following year, he and Paolo Venini, a Murano glass producer with progressive tastes, conceived an unusual vase with a highly distinctive shape. With its striking folds and points that resembled a fluttering handkerchief (*fazzoletto* in Italian), the Fazzoletto vase quickly became a bestseller.

To make the vase, a square sheet of malleable, hot glass was manipulated into shape inside a mould. The first vases were made in plain coloured glass, but the glassworks was soon producing the vases in a huge variety of different colours, patterns, and sizes. In the years that followed, other manufacturers brought out similar designs. The unusual shapes caught people's imagination and they were seen not just as flower vases but as colourful sculpture that would adorn a sideboard or table. The Fazzoletto and its imitators were very popular in the 1950s and 1960s, and few owners realized that the original vase was the work of an innovative Italian designer in collaboration with a Venetian glass-maker.

The raised points resemble the corners of a handkerchief

### PAOLO **VENINI**, FULVIO **BIANCONI**

**1895-1959, 1915-96**

Paolo Venini began his working life as a lawyer, but in the 1920s he bought a glassworks on Murano and quickly established himself as a leading manufacturer of glass. He soon had a reputation for employing talented artists to produce innovative products – early collaborators were the leading architects and designers Gio Ponti and Carlo Scarpa.

▲ **Paolo Venini**

Fulvio Bianconi was apprenticed at the Murano glassworks for a year in 1931, at the age of 16. He went on to become a successful graphic designer, working with several Italian companies, most notably the publisher Garzanti. After the success of the Fazzoletto vase, he combined graphics work with designing glass. Often working alongside the master glass-makers at the furnace, Bianconi explored new colours and forms, and sometimes even moulded the objects himself.

▲ **Fulvio Bianconi**

# Visual tour

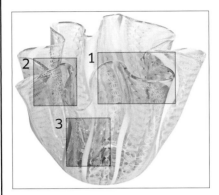

**KEY**

◀ **EDGE** The edge of the vase combines a series of curves with different heights, creating a sense of continuous movement. Where the curved edge doubles back, there are two layers of glass to look through. These folds in the vase catch the light, emphasizing the delicate patterns and colours.

1

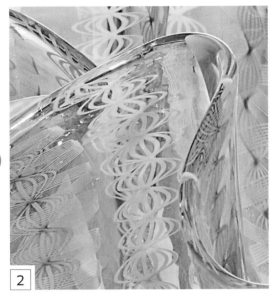

2

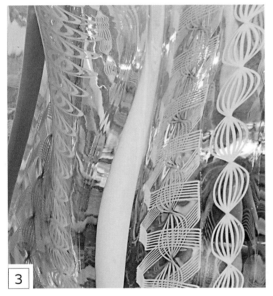

3

▲ **POINT** The glass comes to a fold at irregular intervals. If the vase is held upside down, these folds resemble the corners of a handkerchief hanging down. They give the piece a distinctive organic profile, making the Fazzoletto vase stand out from other glassware of the period.

▲ **SURFACE** The walls of the vase are made up of a series of gentle curves and deep undulations that enhance the complex, lacy frills of the filigree. They follow the line of the original glass mould and are designed to look like folded and draped cloth. The Fazzoletto vase has been produced in numerous patterns, sizes, and colours.

This is a fine example of filigree glass: glass with white or coloured glass threads that are incorporated into a clear body

## ON **DESIGN**

The glass-makers of Murano had used the same techniques for centuries and many of their designs derived from Roman forms probably adopted when glass was first made on the island in the 13th century. Paolo Venini changed this approach by hiring leading designers and encouraging them to develop their own use of pattern and colour. Among Bianconi's most celebrated designs for Venini were the Pezzato (Patchwork) vases, in which rectangles of coloured glass were fused onto clear glass to create vibrant patterns. Using another technique, known as murrine, he set identical slices from rods of multicoloured glass onto a plain background. In his bold designs, Bianconi combined bright colours with a strong sense of shape and form.

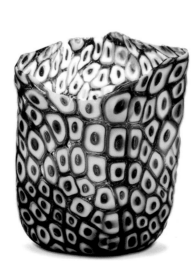

▲ **Pezzato vase**          ▲ **Murrine vase**

# Atomic wall clock

1949 ▪ PRODUCT DESIGN ▪ WOOD AND METAL ▪ USA

SCALE

## GEORGE NELSON

**For several hundred years**, nearly all clock faces were essentially the same: they consisted of a disc bearing Roman or Arabic numerals and a pair of simple pointed hands. George Nelson changed this by rethinking clock design. In his innovative Atomic clock, the twelve small spheres that represent the hours resemble the balls used in scientific models of atoms and molecules, and this

electric-powered clock reflected the contemporary fascination with science, progress, and modernity. Nelson said that he was not entirely sure who came up with the original idea. He described a sociable evening with a group of his designer friends, Isamu Noguchi (see p.104), Buckminster Fuller, and Irving Harper. The four men sat together drinking and passing around a piece of drafting paper, on which each drew different designs for clocks. When Nelson looked at the paper the next morning, he could not remember who had done which sketch, but he was sure that the Atomic clock was the one to develop. Enthusiastically received by the Howard Miller Clock Company, it soon became an icon of modern design and, together with other influential pieces, established Nelson as a significant figure in American Moderrism.

### GEORGE **NELSON**

#### 1908-86

George Nelson studied architecture at Yale and in Europe, where he discovered the Modernism of Le Corbusier and his followers. Back in the USA, he combined the practice of architecture with writing about building and design. His ideas caught the eye of D.J. De Prée, President of Herman Miller, who hired Nelson as director of design. Nelson was keen to nurture new designers, and their work for Herman Miller, together with Nelson's own innovative furniture, helped introduce countless Americans to modern design.

# Visual tour

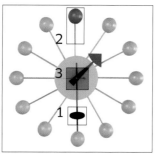

**KEY**

▶ **HOUR SPHERE** The spheres that are used to mark the hours are made of beautifully finished polished wood. Like the balls used in atomic modelling kits, they are held on thin metal rods that fit tightly into a hole in each sphere.

▼ **CENTRE**
The hour hand is broadest at the end that is attached to the centre of the clock, to make it easier to connect to the mechanism beneath. The join is finished with a round brass cap that matches the metal rods supporting the spheres.

▶ **MINUTE HAND**
A large black ellipse on the minute hand gives it the appearance of a metronome. It catches the eye, making it easier to read the time, and adds a quirky touch of humour.

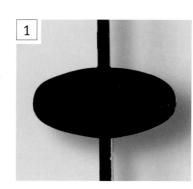

1

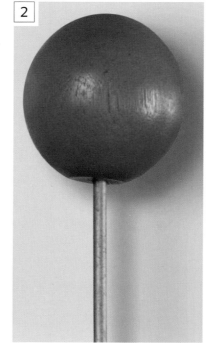

2

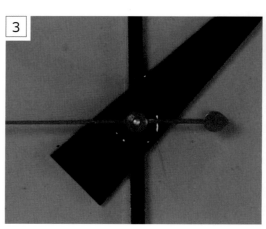

3

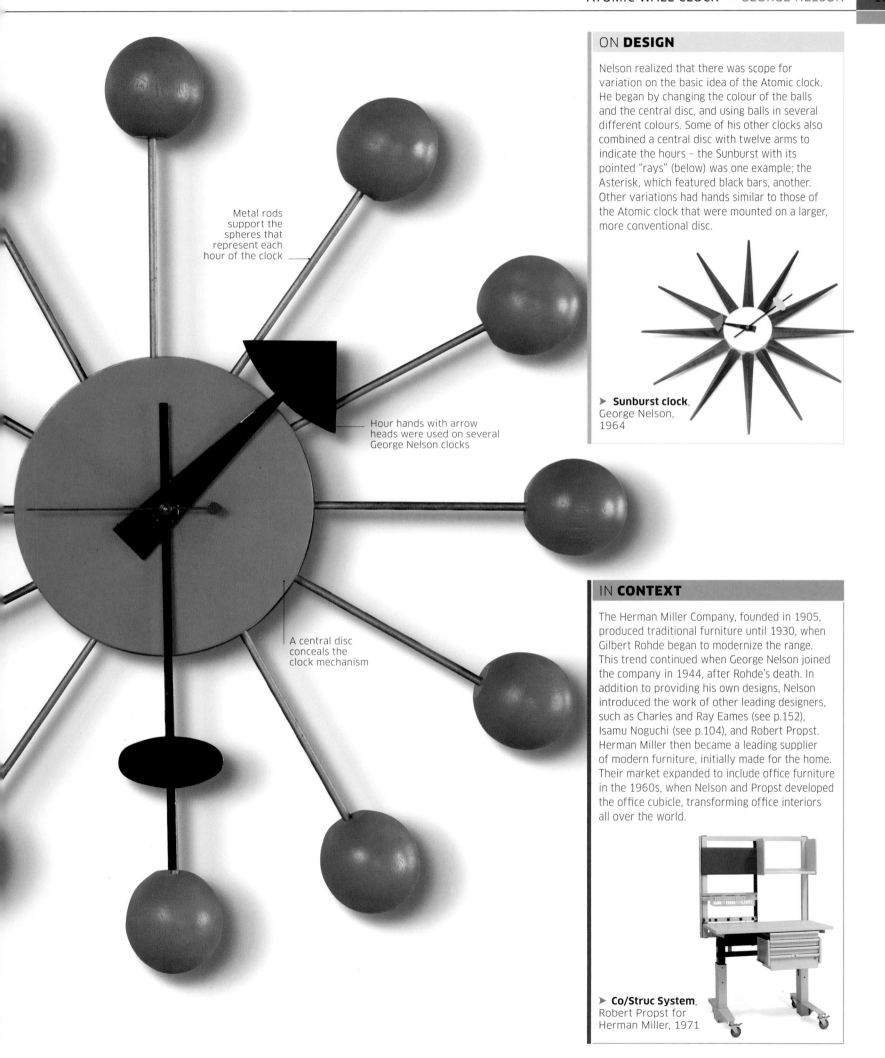

Metal rods support the spheres that represent each hour of the clock

Hour hands with arrow heads were used on several George Nelson clocks

A central disc conceals the clock mechanism

## ON **DESIGN**

Nelson realized that there was scope for variation on the basic idea of the Atomic clock. He began by changing the colour of the balls and the central disc, and using balls in several different colours. Some of his other clocks also combined a central disc with twelve arms to indicate the hours – the Sunburst with its pointed "rays" (below) was one example; the Asterisk, which featured black bars, another. Other variations had hands similar to those of the Atomic clock that were mounted on a larger, more conventional disc.

▶ **Sunburst clock**, George Nelson, 1964

## IN **CONTEXT**

The Herman Miller Company, founded in 1905, produced traditional furniture until 1930, when Gilbert Rohde began to modernize the range. This trend continued when George Nelson joined the company in 1944, after Rohde's death. In addition to providing his own designs, Nelson introduced the work of other leading designers, such as Charles and Ray Eames (see p.152), Isamu Noguchi (see p.104), and Robert Propst. Herman Miller then became a leading supplier of modern furniture, initially made for the home. Their market expanded to include office furniture in the 1960s, when Nelson and Propst developed the office cubicle, transforming office interiors all over the world.

▶ **Co/Struc System**, Robert Propst for Herman Miller, 1971

# Calyx furnishing fabric

1951 ▪ TEXTILE DESIGN ▪ PRINTED LINEN ▪ UK

SCALE

## LUCIENNE DAY

**The avant-garde textile design Calyx** became one of the most familiar fabric patterns of the 1950s, summing up for many the combination of strong colours, bold graphics, playfulness, and optimism that typified the style known as mid-century Modern. With its variety of abstract, plant-like shapes and spidery linear elements, Calyx looked strikingly different from other textiles on sale in 1951. Refreshing in its combination of contemporary, stylized forms and a distinctive yet restrained colour palette, the pattern had an upbeat feel. Like other Modern designs produced at around the same time, it seemed to symbolize the UK's hopes for recovery in the aftermath of World War II.

Calyx was the work of pioneering British designer Lucienne Day. When she showed the design to London furnishers Heal's, they were sceptical, thinking it too radical to compete in a market dominated by conventional floral prints. Tom Worthington, Director of Fabrics at Heal's, paid Day half the usual fee because he thought he would never manage to sell any of the fabric. When Calyx was displayed in the popular Homes and Gardens Pavilion designed by Lucienne's husband Robin Day for the 1951 Festival of Britain, however, it was favourably received.

Later the same year, the fabric won a gold medal at the prestigious Milan Triennale in Italy, and a major award from the American Institute of Decorators followed a year after that. Calyx was suddenly a commercial success. Heal's paid Day the remainder of her fee, and she followed Calyx with several other fabric designs in a similar style. Her curtain and upholstery fabrics were soon seen wherever people wanted to make their homes look bright, sophisticated, and modern.

> "In the very few years since the end of the war, a new style of furnishing fabrics has emerged"
>
> **LUCIENNE DAY**

### LUCIENNE **DAY**

#### 1917–2010

Désirée Lucienne Conradi was born in Surrey, UK, to a British mother and a Belgian father. She went to art school in Croydon and then attended the Royal College of Art, where she met her future husband, fellow designer Robin Day. She taught during World War II and became a freelance textile designer in 1946, starting with dress fabrics but quickly moving on to furnishing fabrics. By the late 1940s, Day was selling designs to Heal's, who purchased Calyx, her breakthrough design. She produced around 70 designs for Heal's, as well as many for other manufacturers, plus an array of table linen, wallpapers, and carpets. Her work on carpets (both as designer and as colour consultant) was particularly successful, leading to a Design Centre Award. Commissions from overseas clients followed, including one from the German manufacturer, Rosenthal, for whom she designed tableware. In the economic downturn of the 1970s, Day stopped her industrial design work and concentrated on a series of large-scale, one-off tapestries, which she created from colourful pieces of silk. These beautiful, abstract "silk mosaics" were widely exhibited.

# Visual tour

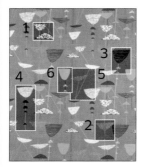

**KEY**

> **WHITE FORMS** Many of the shapes that make up the Calyx design are pale – often predominantly white – with stippled decoration in shades of brown, black, and yellow. This sophisticated treatment adds form and texture, and the pale coloured shapes seem to float out of the earthy, olive tones of the background. The use of pale colours with stippling was much imitated and became a hallmark of 1950s design.

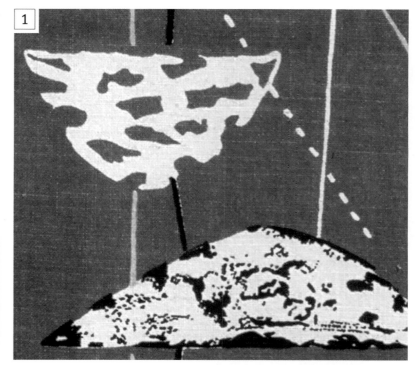

1

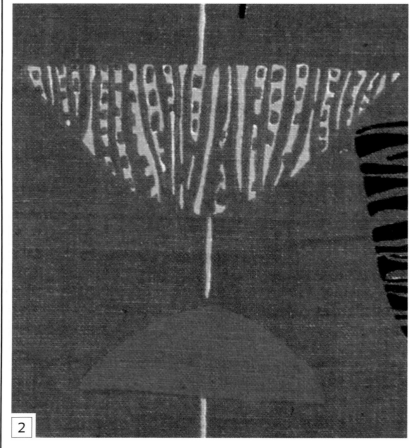

2

▲ **ABSTRACT SHAPES** The repeated organic forms in the pattern are reminiscent of mushroom caps, plant seedheads, and flowers in profile. By naming the design Calyx, a botanical term used for the protective "cup" around a flower bud, Day was confirming that these shapes were stylized, abstracted plant forms.

> **DARK FORM** One of the strongest elements in the pattern is a dark shape made up of wavy, horizontal lines in black, white, and brown. The lines seem to follow the weave of the linen fabric and add a textural element that gives the flat design depth. The use of the repeating image leads one's eye across the pattern.

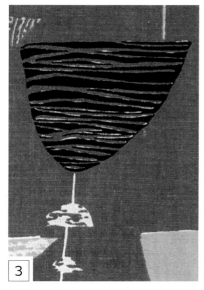

3

4

▲ **BLACK AND YELLOW MOTIFS** In this part of the pattern, there are finely drawn, elongated yellow lines punctuated by smaller, parasol-like motifs in black. Unlike many of the realistic floral designs of the period, the forms in Day's design are not naturalistic. With the deftest of touches, Day skilfully suggests the slender, growing stems of plants.

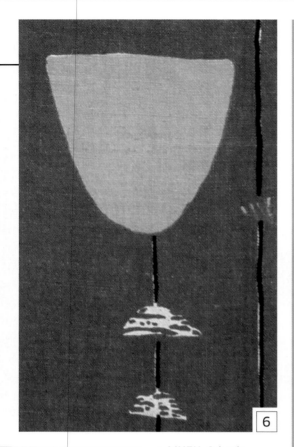

## ON **DESIGN**

The success of Calyx, both at the Festival of Britain and elsewhere, did much to popularize Lucienne Day's textile designs at home, as well as internationally. Her fine-art inspired patterns, which reveal the influence of contemporary Surrealist and Modern movement painters, began to appeal to a wide audience. As a result, she was busy throughout the 1950s and 1960s, creating new fabrics for a variety of clients. Day attributed her success to a reaction against the dreariness and lack of consumer choice during World War II. Some of her fabrics were wholly abstract, made up of a range of geometrical shapes like the triangles in Isosceles (below right). Others, such as Herb Antony (below left), developed her use of semi-abstract plant forms, with bold outlines of stylized flowers, leaves, and roots reminiscent of the images of the Catalan painter Joan Miró. Day's innovative and confident use of colour also set her designs apart. In Herb Antony, the skeletal, white outlines of plants on a black background are enlivened with bright splashes of primary colour. Isosceles is a more formal, linear composition that combines vivid turquoise, yellow, and black.

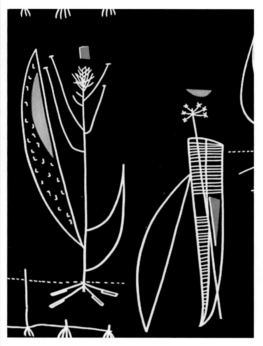

▲ **Herb Antony**, 1956

▲ **Isosceles**, 1955

◄ **LINEN** Calyx is screen-printed on linen, and the weave of the fabric is clearly visible, particularly in the olive background and where the pattern consists of blocks of plain colour. Using linen made Calyx quite an expensive fabric, so Day followed it in 1952 with Flotilla. This similar pattern was printed on rayon and had a smaller repeat, which made it much more affordable.

◄ **LINEAR ELEMENTS** Acid-yellow, black, and white vertical and diagonal lines run all the way up and down its pattern, linking the organic shapes like stalks. Day introduced variety by making some of the lines continuous and others broken, but they are all strong graphic elements that give the design unity.

## IN **CONTEXT**

Following a visit to Scandinavia in the late 1940s, Lucienne and Robin Day were keen to promote a strong, modern design movement in the UK. At the Festival of Britain in 1951, Calyx was displayed alongside steel and plywood furniture by Robin, and the couple hoped that people would be inspired by the exhibition to transform their homes. The couple's interior schemes for their London flat and then their large house in Chelsea were featured in many fashionable magazines, and they were compared with Charles and Ray Eames (see p.152). When working for commercial clients, however, the Days usually worked independently – Robin designing furniture and Lucienne concentrating on fabrics. Perhaps the most successful of their occasional collaborations was the 1960s cabin interior for the BOAC Super VC10 aircraft, which featured Robin's comfortable seating alongside Lucienne's chic, modern decoration.

▲ **BOAC aircraft interior** combining the work of Lucienne and Robin Day

# FESTIVAL OF BRITAIN

## MAY 3 – SEPTEMBER 30

MAIN INFORMATION CENTRE · SWAN & EDGAR BUILDING · PICCADILLY CIRCUS · LONDON

Printed in Great Britain for H.M. Stationery Office by W. S. Cowell Ltd, Ipswich and London.

# Festival of Britain symbol

1951 ■ GRAPHICS ■ COLOUR LITHOGRAPH ON PAPER ■ UK

SCALE

## ABRAM GAMES

**In the summer of 1951, a landmark festival** was held in the UK, with exhibitions and other events at a variety of locations around the country. Designed to be a morale-booster in the austere postwar years, showcasing national achievements in every area of art and science, the festival promoted its events in a distinctive graphic style, brilliantly captured by Abram Games's celebrated symbol.

Games was one of twelve designers invited to take part in a competition for a symbol to represent all the activities of the Festival of Britain. The brief required that it should be "simple in design, recognizable at a glance", and effective at a wide range of sizes. Games's entry combined compass points with a stylized head of Britannia and the year of the festival, 1951, in italic numerals. He submitted versions in both black and white and in colour (red, white, and blue). The committee judging the competition selected Games's entry, but asked the designer to make the symbol more festive. Inspired by watching his wife hanging out washing on a windy day, Games added a swag of bunting to the symbol. The adapted motif was used widely on

everything from architectural signs to postage stamps, and appeared on official festival guide covers designed by Games and on posters (such as the one pictured, left) created by others. The symbol's strong graphic form was instantly recognized, and its combination of gravitas and festivity made it a fitting icon for the festival.

### ABRAM **GAMES**

#### 1914-96

After briefly attending art college in his native London, Abram Games was employed as an office boy by a commercial art studio. He set up as a freelance graphic designer in the 1930s, producing posters and publicity material for many high-profile clients, including Shell, BP, Guinness, and the Post Office. During World War II, Games was an Official War Artist, and created the famous "Blonde Bombshell" recruitment poster for the Auxiliary Territorial Service (ATS). His Festival of Britain work is his most celebrated achievement.

# Visual tour

**KEY**

▶ **FESTIVAL SYMBOL**
With a carefully formed profile and the shadow cast by the helmet, Games's depiction of the head of Britannia is simple but effective. The beak-like visor of her helmet echoes the strong line of her angular nose.

1

2

▲ **1951 TYPEFACE** Including the year provided a reminder that the festival took place a hundred years after the Great Exhibition of 1851. Games chose an italic typeface with a strong shadow. This style of typography, using bold letters, often with shadows, was widely used during the festival.

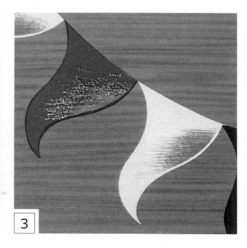

3

▲ **BUNTING** Games gave a fluttering quality to the bunting by putting a curve into the long edges of the flags. A simple shadow effect was achieved in this poster (produced by an unknown printer at the London Press Exchange) by allowing patches of the woodgrain background to show through.

# Birch platter

1951 ▪ TABLEWARE ▪ BIRCH PLYWOOD ▪ FINLAND

## TAPIO WIRKKALA

SCALE

Tapio Wirkkala, one of Finland's best-known 20th-century designers, worked on many different kinds of object in a great variety of materials. He always searched for a more organic approach to modern design, and his pieces, which include glassware shaped like mushrooms or blocks of melting ice, often depict natural forms. Fundamentally, he wished to exploit the nature of the materials he used to best effect. The finest examples of this approach include the series of plywood sculptures and objects that he made in the early 1950s.

Plywood is made of thin sheets of wood glued together, with the grain of each layer usually lying at right angles to those of the layers above and below it to increase strength and minimize warping. Wirkkala made his own

plywood, but instead of using it in sheets, he built up numerous layers to create a large block. He then cut into it to expose the edges of the layers, which displayed a series of stripes of different shades and thicknesses. The contrast between the pale wood and the darker layers of glue, together with the variation in the width of the layers according to the angle of the cut, created an effect similar to that of the natural annual rings in a tree trunk. By skilfully exploiting the curves and contours produced from the layers of plywood, Wirkkala aimed to create objects that had a unique organic quality. With its warm tones and beautifully balanced forms, this birch plywood platter reveals the designer's deep respect for natural materials and his highly developed sculptural sensibility.

### TAPIO **WIRKKALA**

#### 1915-85

After training as a sculptor in Helsinki, Tapio Wirkkala became a successful designer. From 1947 until his death, he directed design at the Iittala glassworks, while also proving his versatility in a freelance capacity. His designs included glassware for Venini, ceramics for Rosenthal, a light bulb for Aram, and his own version of the traditional Finnish sheath knife (one of which he used for carving) for Hackman. Wirkkala was also responsible for the Finnish pavilions at the 1952 and 1954 Milan Triennali. His innovative work helped to establish Finland's international reputation for design excellence.

"All materials have their own unwritten laws... the designer should aim at being in harmony with his material"

**TAPIO WIRKKALA**

The platter has a smooth, perfectly finished surface

# Visual tour

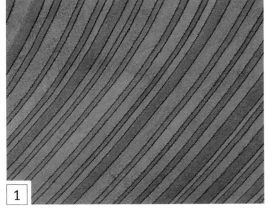

## KEY

▶ **WOOD PATTERN** Wirkkala's platter derives much of its character from the pattern created by the light and dark brown layers in the plywood. The layers are not uniform, as they are in standard industrial plywood, but vary in thickness, giving the platter a natural, organic quality.

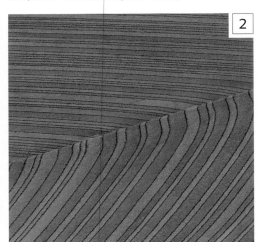

2

1

◀ **CENTRAL CREASE** Where the curved layers of wood bend back on themselves at the centre of the platter, the carved surface forms a ridge or crease, similar to the central rib of a leaf.

▼ **EDGE** The edge of the platter is smooth, creating a gradual curve. Like all Wirkkala's wooden pieces, the birch platter has a tactile quality, which makes it a pleasure to handle and use.

3

## ON **DESIGN**

Tapio Wirkkala is well known for his wide-ranging glass designs. His work from the 1940s and 1950s includes a number of glass vessels produced by the renowned Finnish manufacturer, Iittala. Among these pieces are vases and dishes engraved with fine lines similar in appearance to the veins of a mushroom or a leaf and like the "veins" in the designer's plywood bowls. Other glassware ranges from beautifully simple glasses and goblets, such as Iittala's Tapio range, to the much imitated Ultima Thule range, which has frosted surfaces that look like ice. Wirkkala employed a variety of techniques to achieve magical effects in his glassware, from engraving and acid-etching to the ingenious use of wooden moulds that burned, altering their surface texture, on contact with the molten glass.

▲ **Turned Leaf** glass dish, Tapio Wirkkala, 1953

The layers were bent during production to create a distinctive striped effect

# Kilta tableware

1952 ▪ CERAMICS ▪ GLAZED EARTHENWARE ▪ FINLAND

## KAJ FRANCK

**In the late 1940s**, Kaj Franck, head of the design studio at the Finnish Arabia ceramics company, began to create a new collection of tableware. Knowing that in the postwar period, many Finns lived in small apartments, few people had domestic help, and more and more women were going out to work, Franck realized that the traditional dinner service, with its numerous decorated items made of delicate porcelain, was no longer appropriate. In its place, he created a plain-coloured range of more than 30 "mix-and-match" pieces in robust, glazed earthenware. The items were designed to stack, making them easier to store, and some were dual-purpose. Franck based his designs on geometric shapes – circles, cones, and cylinders – and kept the lines clean and simple. The colours were restricted to white, black, yellow, green, and blue.

The plain, simple crockery was so different from what people were used to that sales were initially poor, but when Arabia organized a series of demonstrations, touring exhibitions, and talks, people became more enthusiastic and realized that the inexpensive tableware was elegant, practical, and ideal for everyday use. Kilta began to win prizes, then became a bestseller, and came to be regarded as a symbol of the flair and practicality of Finnish design. Although Kilta was discontinued in the 1970s, Franck later designed an updated form in vitreous china, called Teema. The set was relaunched in 1981 and is still sold by Iittala, one of the leading Finnish design companies – clear affirmation of the design's enduring appeal.

### KAJ **FRANCK**

#### 1911–89

After studying furniture design in Helsinki, Franck designed lighting and furnishings in Copenhagen. In 1945, he returned to Finland to join the Arabia pottery and porcelain factory, where he was head of the design department until 1961. Franck also produced glass designs for Iittala. From 1950 to 1973 he was Art Director of the Nuutajärvi-Notsjö glassworks, becoming known as a designer who focused on essentials. Franck was also a noted teacher and became artistic director at Helsinki's College of Applied Arts, where he had studied.

## Visual tour

**KEY**

▲ **PITCHER LIP** The details in the Kilta range are simple and understated, like this small lip on the pitcher – in fact, the lip is large enough to work well. The original pitcher had a cork stopper so that milk or cream could be stored in it.

◀ **CUP HANDLE** The handles on the Kilta cups, pots, and other items are strong, with rounded edges, and are big enough to grip comfortably. They are a far cry from the small, fragile handles seen on many cups before World War II.

▶ **KNOB** The Kilta lids have a small, circular knob, slightly flared towards the top to make them easy to pick up. The dimple on top of the knob adds visual interest, but does not compromise the simplicity and geometry of the overall design.

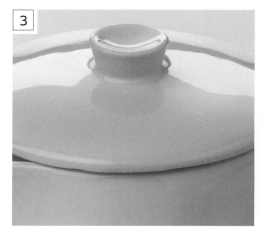

# "Is it not the ultimate meaning of beautiful to be essential, functional, justified, truthful?"

**KAJ FRANCK**

In 1950, the Finnish engineering company Wärtsilä, which already owned the Arabia pottery, bought the historic Nuutajärvi-Notsjö glassworks in southwestern Finland. Kaj Franck became Art Director of the glassworks in 1951 and did some of his best work for them. His glassware had the same simplicity of form and elegant lines as his ceramics for Arabia. His colour palette included delicate shades of smoked glass and he sometimes added decorative effects, such as silvered lustre, for an opulent finish. Franck also used cased glass, in which a vessel is made of more than one layer of glass, to create surfaces with subtle colours.

▶ **Cased glass vase**, Kaj Franck for Nuutajärvi-Notsjö, 1950s

▶ **Pitcher**

▼ **Teapot**

▶ **Teacup and saucer**

▲ **Creamer with lid**

The saucer could also be used as a small dish

# Diamond armchair

1952 ▪ FURNITURE ▪ STEEL-WIRE MESH ▪ USA

## HARRY BERTOIA

SCALE

**In the 1940s, sculptor Harry Bertoia** worked with Charles and Ray Eames (see pp.152–55) on their furniture designs, using his metalworking skills to help them develop metal legs and frames for some of their chairs. Bertoia was, however, disappointed not to receive credit for his work, so he left to make a series of chairs, including his famous Diamond armchair, for the Knoll furniture company. Using wire mesh to create an impression of lightness and transparency, he designed chairs with organic forms that followed the contours of the human body.

Bertoia realized that wire welded together to form a gridded mesh was strong, so he could use relatively thin material to create a rigid, contoured shell for the seat,

back, and arms of the chair. Viewed from the front, the entire shell has a diamond shape, and the glistening mesh is made up of several, smaller "diamonds", giving the whole chair a pleasing visual coherence. Thicker metal rods were required to create a strong base, which Bertoia splayed out at the front to mirror the broadly swelling form of the chair's arms. Knoll sold the chair with a specially shaped cushion, but the full effect of the elegant lattice of diamonds, and the way the shapes seem to change as you walk around the chair, are best appreciated without it. As Bertoia said of his wire-mesh chairs, "...they are mostly made of air, just like sculpture. Space passes right through them."

## HARRY **BERTOIA**

### 1915–78

Italian-born Harry Bertoia settled in the United States in 1930 and went to art school in Detroit and at Cranbrook Academy of Art, Michigan. He worked as a jeweller and metalworker, teaching metalwork to students at Cranbrook before collaborating with Charles and Ray Eames on their furniture projects. When he was commissioned by Knoll, the company provided him with his own studio space, in which he developed his range of wire-mesh chairs. The payment he received for the designs enabled Bertoia to buy a house and workshop in Pennsylvania, where he devoted himself to sculpture for the rest of his life. A great music lover, his work includes many sound sculptures composed of thin metal rods.

▲ **Harry Bertoia** in his Pennsylvania studio, 1956

# Visual tour

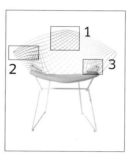

**KEY**

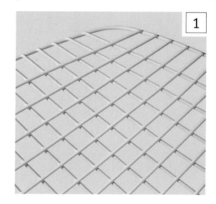

| 1 | ◄ **CHAIR BACK** The distinctive diamond pattern of the wire mesh is fairly regular, and manages to convey the impression of strength combined with delicacy. Although the latticework shell might appear to be hard and unyielding, the rods have been carefully shaped to offer maximum comfort.

**2**

▲ **CURVE AND ARM** The diamonds change shape as they curve to form the armrests, making them look more irregular than those on the back. Their reflective chrome surfaces catch the light, creating interesting shapes and shadows.

▶ **JOINT** Bertoia originally designed a rubber joint to link the shell and support. It was later replaced with this metal track that holds both parts together with screws. Visually, this was a better solution, and metal was also more hard-wearing than rubber.

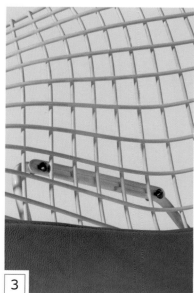

**3**

# "The urge for good design is the same as the urge to go on living"

**HARRY BERTOIA**

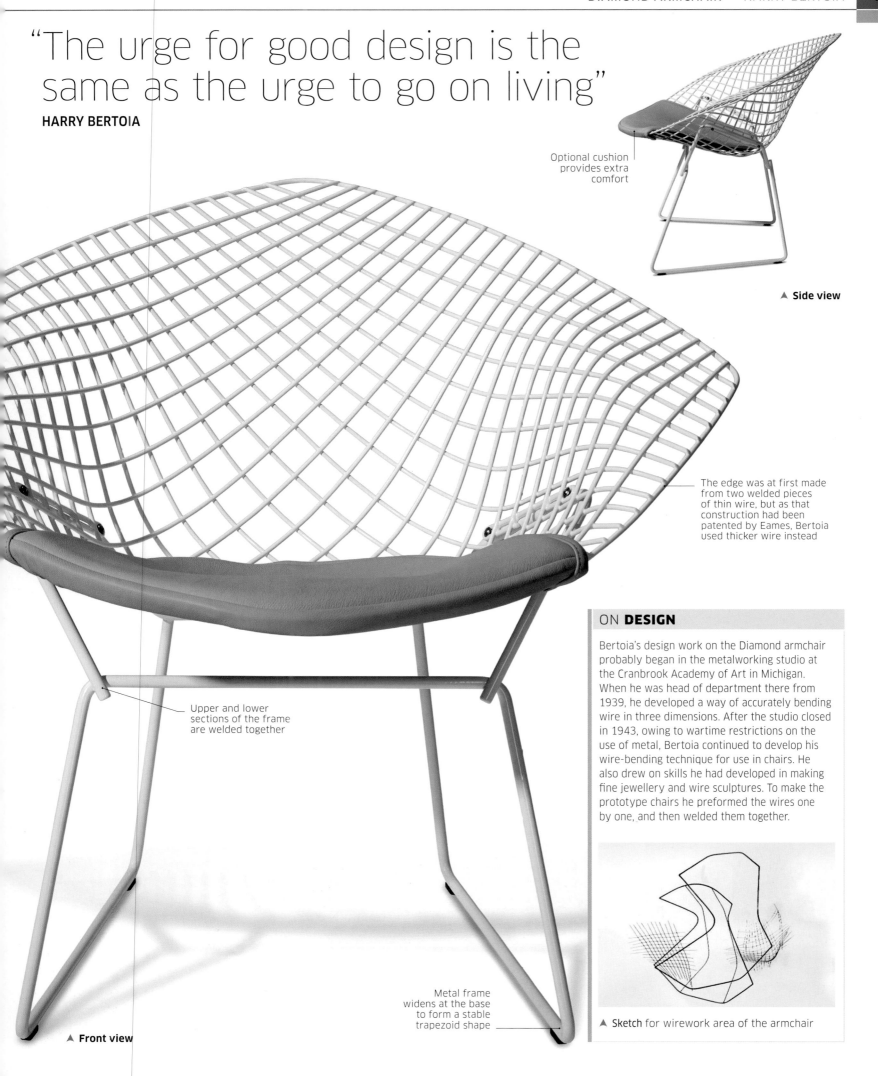

Optional cushion provides extra comfort

▲ **Side view**

The edge was at first made from two welded pieces of thin wire, but as that construction had been patented by Eames, Bertoia used thicker wire instead

Upper and lower sections of the frame are welded together

Metal frame widens at the base to form a stable trapezoid shape

▲ **Front view**

## ON **DESIGN**

Bertoia's design work on the Diamond armchair probably began in the metalworking studio at the Cranbrook Academy of Art in Michigan. When he was head of department there from 1939, he developed a way of accurately bending wire in three dimensions. After the studio closed in 1943, owing to wartime restrictions on the use of metal, Bertoia continued to develop his wire-bending technique for use in chairs. He also drew on skills he had developed in making fine jewellery and wire sculptures. To make the prototype chairs he preformed the wires one by one, and then welded them together.

▲ **Sketch** for wirework area of the armchair

# Pride cutlery

1953 ■ FLATWARE ■ SILVER PLATE ■ UK

SCALE

## DAVID MELLOR

**There is a long history of cutlery production in the UK**, stretching back to the 18th century. Many of the manufacturers were based in Sheffield, an internationally renowned centre of steel production and metalworking, and achieved a very high standard of workmanship. Their designs, however, tended to be conservative. In 1953, Sheffield-born David Mellor, a student at London's Royal College of Art, created a set of cutlery as part of an art-school project that respected tradition while also embracing modernity. The design, which he called Pride, embodied the high-quality, immaculate finish, and elegant proportions of traditional cutlery, but had sleek, modern lines and no ornament. With its long-tined forks, oval-bowled spoons, and knives with resin handles, the flatware looked good in any setting, from a Georgian dining room to a fashionable 1950s restaurant.

One of Mellor's fellow students at the Royal College was Peter Inchbold, whose family owned Walker and Hall, a firm of Sheffield silversmiths. Inchbold suggested that the company produce Pride, and in 1954, they not only agreed but also appointed the newly graduated Mellor their design consultant, which at that time was an unusual role in British industry. Pride was influential in showing the European cutlery industry that modern style could be combined with traditional values to produce an enduring design.

### DAVID **MELLOR**

#### 1930–2009

After the success of Pride, British metalworker and designer David Mellor continued to create cutlery for Walker and Hall, including flatware in stainless steel. He was commissioned by the UK government to design the Embassy range (cutlery and a tea pot) for British embassies around the world, and the Thrift stainless-steel range for state canteens, hospitals, and prisons. In the 1970s, Mellor began to manufacture and sell his own flatware, concentrating solely on cutlery until the late 1990s, when he also undertook other design work.

# Visual tour

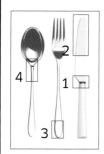

**KEY**

**► KNIFE BLADE**
The blade has a straight back but the edge curves, very gently and almost unnoticeably, towards the top. The tip of the blade, like that of most traditional table-knife blades, is rounded, to avoid unnecessary sharp points.

**► HANDLE AND JOINT**
The clear break between the handle and the knife blade is emphasized by the change in material from metal to acetal resin, which recalls traditional ivory or bone handles. In a modern touch, however, the blade has been cut at a steep, 45-degree angle to further reinforce the distinction.

 1

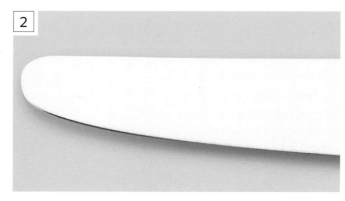 2

3

**▲ END OF FORK HANDLE** The ends of the metal handles are beautifully designed as curves that do not quite come to a point. The resulting shape, which avoids sharp edges and lacks the decoration of traditional cutlery, is both comfortable to hold and elegant.

**► SPOON HANDLE** This section of the handle, like that of the fork, is very slender, yet thick enough to hold. The unusually slim profile of both the spoon and fork handles gives the pieces a sleek, modern appearance.

 4

# "...choosing what we use, choosing how to live"

**DAVID MELLOR** on being a designer

Although David Mellor produced some widely used designs for street furniture (seating, bus shelters, and traffic signals) in the UK, he concentrated mainly on cutlery. From the time he began manufacturing, David Mellor responded to fashion, producing designs from the informal Chinese Ivory range with resin handles of the 1970s (below) with its straight knives and forks, to Minimal, the pared-down, flat set that he created in 2002. Although stylish, each of these cutlery designs is very carefully considered, and Mellor never lost sight of the need for balance and elegance in a knife, fork, or spoon, however simple the design may appear. Mellor's cutlery and kitchenware is still made in the Round Building, Mellor's purpose-built factory in Derbyshire, now overseen by his designer son, Corin.

▲ **Chinese Ivory cutlery**, David Mellor, 1977

In the second half of the 20th century, stainless steel was adopted more widely for cutlery. David Mellor proved that it could be used to produce well-designed cutlery for a premium market, and other mass-market manufacturers soon followed suit. The most astute employed talented designers to create items that could be mass produced. One such company was Viners of Sheffield, who commissioned designs such as Studio and Shape from the renowned British goldsmith Gerald Benney. Like Mellor, Benney drew on a range of influences, from the pared-down work of Scandinavian designers to the craft traditions of British metalwork, but he was more willing than Mellor to add decorative patterns to the handles of his knives, forks, and spoons.

▲ **Studio cutlery** for Viners of Sheffield, Gerald Benney, 1960s

The immaculate surface of the spoon reveals the high level of hand-finishing

The swelling base of the handle helps to balance the tines

The acetal resin handle was produced in either white or black

▲ **Spoon**          ▲ **Fork**          ▲ **Knife**

# Fender Stratocaster

1954 ▪ PRODUCT DESIGN ▪ WOOD, PLASTIC, AND METAL ▪ USA

SCALE

## LEO FENDER

**In 1954, an electric guitar appeared on the market** that transformed popular music. The radical shape and unique, bright tone of the Stratocaster, affectionately known as the "Strat", gave musicians a new expressive power. During the 1960s, it was played by more and more guitarists, and the sounds and techniques they developed took popular music, from country to rock 'n' roll, in exciting new directions. With its innovative, horned body shape, Leo Fender's design was truly attention-grabbing – this,

combined with its cutting-edge sound, meant that the Stratocaster was a major factor in the emergence of a new phenomenon in popular music: the guitar hero.

Leo Fender had enjoyed successes in the early 1950s with his Telecaster, the first mass-produced solid-bodied electric guitar, and his Precision Bass, the first electric bass guitar to be widely played. Fender did not play himself, so when he embarked on plans to create another guitar, he took advice from musicians such as Bill Carson,

The body is beautifully contoured, with chamfered edges and an organic, curved shape that makes it supremely comfortable to play

The strap attaches to metal buttons

Ash body with a two-tone sunburst finish

Pick-ups

The tremolo arm (removed) attaches to the bridge

Tone controls

Volume control

The scratchplate protects the wood of the body and covers the electronics

Curves and cutaways allow easy accessibility to the highest notes the Western swing guitarist. Carson and his colleagues wanted a guitar that was light, comfortable to play, and had a good sound; some also wanted an instrument that was easy to maintain and customize. Fender responded to all these needs. He built a guitar with smooth edges and a curving shape that moulded itself to the musician's body, making it easy for the player to reach all the frets. The instrument is beautifully balanced and its signature features – three pick-ups and a tremolo arm – gave an impressive range of sounds. The bolt-on neck was easier and cheaper to produce than glued-on "set" necks, and it was simple to repair, modify, or replace.

Although the guitar's unusual shape took some players by surprise, most were won over when they heard the distinctive tone, or tried the instrument for themselves. A breakthrough came in 1957, when the American singer-songwriter Buddy Holly played a Stratocaster on the Ed Sullivan show, introducing millions of people to the first guitar hero and showing them what the instrument could do. Since then, the Stratocaster's popularity has grown and endured. It has been the instrument of choice for some of the world's greatest guitarists, including Jimi Hendrix and Eric Clapton, and its striking, contoured shape and bold finishes have been widely copied.

## LEO **FENDER**

### 1909–91

California-born Leo Fender trained as an accountant and taught himself electronics in his spare time. He started a radio repair business in 1938 and also designed PA systems for musicians. Working in partnership with "Doc" Kauffman, a musician who had also worked for guitar-maker Rickenbacker, Fender began to make electric guitars in the 1940s. The experience led him to take out patents for items such as a guitar pick-up and an amplified lap steel guitar, and to focus his energies on designing an instrument that would be easy to hold, tune, and play. In the 1950s, he created groundbreaking, solid-body electric guitars and Fender became one of the world's most successful guitar brands. Fender sold his company to the CBS corporation in 1965 owing to ill health but when he recovered, he returned to the music business, designing innovative instruments such as the Music Man StingRay bass.

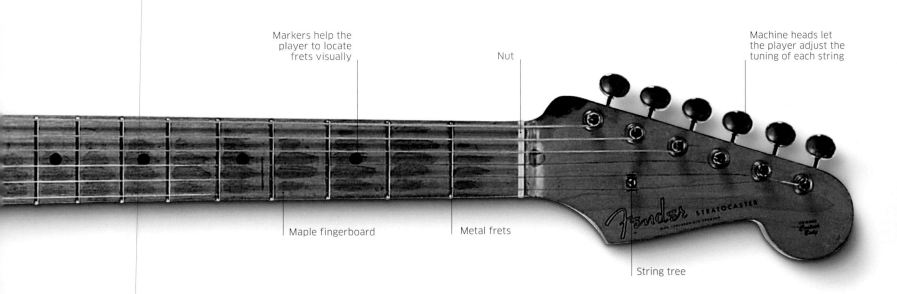

Markers help the player to locate frets visually

Nut

Machine heads let the player adjust the tuning of each string

Maple fingerboard

Metal frets

String tree

## "It was like seeing an instrument from another planet"

**HANK MARVIN**

# Visual tour

**KEY**

▶ **PICK-UPS** The magnetic pick-ups capture the vibrations of the strings and turn them into electrical signals that can be amplified. When Fender designed the Stratocaster, it was usual for electric guitars to have one or two pick-ups. The Stratocaster has three, giving three distinct sounds selected by a switch on the front of the body. The neck and middle pick-ups also have separate tone controls, enabling the player to adjust the sound from high treble to a deeper bass.

▼ **HORN** Fender designed the Stratocaster with distinctive horns, giving a similar shape to his Precision Bass of 1951. The upper horn holds the mounting for the strap. This horned shape with cutaways has a key practical advantage for the player – the cutaways allow easy access to the upper frets (those closest to the bridge), which produce the higher notes.

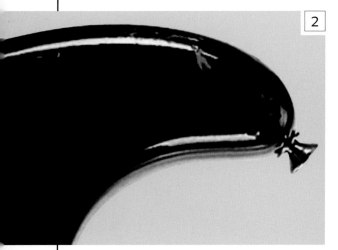

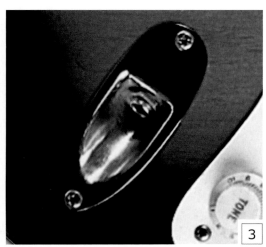

◀ **OUTPUT JACK** The socket that takes the guitar's output jack plug is a metal plate attached to the body with two screws. The rounded plate has a bullet-shaped indentation, so that socket and plug are neatly recessed into the Stratocaster's body, a design that looks good and helps to shield the plug from accidental knocks, both from the front and from below.

▲ **HEADSTOCK** The Stratocaster's distinctive headstock shape is a development of the slimmer version found on its predecessor, the Telecaster. To reduce manufacturing costs, it is flat, rather than angled backwards as on many guitars, so the string tree in the centre is needed to brace the lightest two strings against the nut (the slotted bar at the head end of the fretboard).

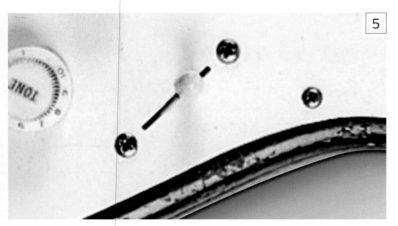

**◄ PICK-UP SELECTOR**

On the original Stratocaster, this switch has three basic positions – one for each pick-up. Players soon found that they could set the switch at the mid-points between the settings, selecting two pick-ups at once and giving two additional, pleasing tones. Later versions have a five-way switch to take advantage of this facility.

5

**► FENDER LOGO**

Thanks to the success of Leo Fender's guitars, especially the Stratocaster, the Fender logo is now recognized all over the world. This early version, which is widely known as the Spaghetti logo, has stylized script letters with thin strokes. The letters are also fairly upright in form. On later versions, the letters have thicker strokes and are more sloping.

6

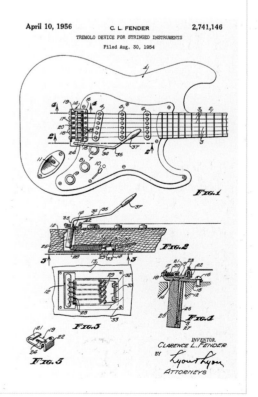

7

**▲ BRIDGE** Each of the guitar's six strings is anchored on its own saddle on the bridge, and each saddle is independently adjustable. By turning a screw, the player can move the bridge saddle slightly, changing the length of the string. This allows a player to fine-tune the guitar's intonation – the accuracy of each note all the way up the fretboard.

## IN **CONTEXT**

The original Stratocaster was made from ash wood with a beautiful black and yellow two-tone sunburst finish. It was a world away from the traditional guitar designs of the time, which sometimes featured carved tops and ornate inlay on the fretboard. Fender, by contrast, relied on the creative use of colour for decoration. Later, the company added red to produce a richer three-colour sunburst effect, and guitars were also produced in a plain blonde, lacquered wood. Further colours followed, included some using car paints in bright red, blue, or green, and others with stunning one-off decoration, such as the Rhinestone (below centre).

**▲ 3-Colour Sunburst**

**▲ Rhinestone**

**▲ Candy Apple Red**

## ON **DESIGN**

Fender filed a patent on a key element of the Stratocaster, the tremolo mechanism, in 1954. It was based on an earlier device, a vibrato arm invented by Paul Bigsby, and it enables the musician to bend the pitch of the strings while playing and return them to their original tuning. It has three main elements – a series of five springs hidden by the backplate, which hold the bridge flat against the guitar body; a tremolo arm; and the bridge itself. By removing two of the springs and adjusting some screws, the bridge can be made to "float". It can be moved by the tremolo arm in either direction, and is held in position by the strings pulling one way and the strings pulling the other. When the bridge is floating, the musician can move the tremolo arm to change the pitch of the notes. Some players prefer to remove the arm and anchor the bridge to a fixed position.

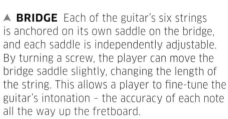

**► Fender patent drawing** for the tremolo device, 1956

# M3 Rangefinder camera

1954 ▪ PRODUCT DESIGN ▪ STEEL, LEATHER ▪ GERMANY

SCALE

## LEICA CAMERA AG

**In 1954, the Leitz company of Germany**, already famous for compact cameras of the finest quality, introduced their latest 35mm model. The Leica M3 raised camera design to a new level by packing an outstanding set of features into a rugged, black-and-silver body. With its immaculately engineered mechanism and high-specification, hand-crafted lens, this Leica set a standard that many other manufacturers tried to follow, but few attained.

Most of the new features of the M3 were designed to make the camera easier to use than its predecessors. Changing lenses could be time-consuming, so a bayonet lens mount was fitted to the M3 to make this much quicker. The camera's clearer, brighter viewfinder was instantly popular with professional photographers and it incorporated a rangefinder system that indicated when the subject was in sharp focus. Another ingenious feature was the frame that automatically appeared in the viewfinder when a longer lens was fitted. This showed exactly how much of the scene would come within the field of view of the lens. The M3's body was nonslip and comfortable to hold, and all the major settings and controls – the lens aperture and distance scales, the shutter-speed dial, the frame counter, the shutter release, and the film

advance – could be seen from above when you held the camera ready to shoot. The M3 was especially popular with photojournalists and travel photographers who were always on the move, so needed the best quality in the most compact package. Leitz manufactured the M3 for more than ten years and many of the innovations were built into later Leicas, and copied by competitors.

Shutter release

Rangefinder window

### LEICA CAMERA AG / ERNST LEITZ GMBH

The original Ernst Leitz optical company (the name was not changed to Leica until 1986) was founded in the 19th century. In 1913, Oskar Barnack, who was responsible for developing the company's microscopes, began to design compact cameras. He realized that a small camera producing a small negative would need high-quality lenses, so that the tiny images would still be sharp when enlarged in the darkroom. Leitz therefore developed a series of lenses specifically for photographic use.

Combining high-quality cameras and lenses proved a winning formula, and the company went on to produce a succession of compact cameras – a line that has continued into the current digital age. Encouraged by the success of their rangefinder focusing, Leica has concentrated on rangefinder cameras and made only a small number of the SLR type embraced by the large Japanese manufacturers. Because Rangefinder models tend to be smaller, lighter, and quieter than SLR cameras, Leica products have found favour with users for whom a compact size is a priority.

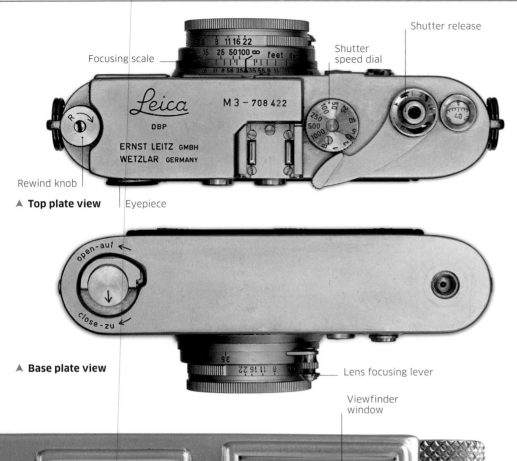

Focusing scale

Shutter speed dial

Shutter release

Rewind knob

▲ **Top plate view**

Eyepiece

▲ **Base plate view**

Lens focusing lever

Viewfinder window

The rewind knob is machined to make it easy to grip

Eyelet for carrying strap

"Shooting with a Leica is like a long tender kiss, like firing an automatic pistol, like an hour on the analyst's couch"

**HENRI CARTIER-BRESSON**

## IN **CONTEXT**

Leicas were designed to take the same kind of small-format 35mm film used in movie cameras, transporting it horizontally through the camera so that each frame would be 24 × 36mm (1 × 1⅓in). The first camera, the Leica I (below), had a long gestation and was eventually launched in 1925. With its curved body, it was recognizably the ancestor of the M3. Over the following decades, the company brought out increasingly sophisticated models, while retaining the compactness and quality that made the first Leica a success. In 1932, the Leica II appeared. This had rangefinder focusing, but a separate viewfinder. The 1933 Leica III added a wider range of shutter speeds and further improvements were made after World War II.

▲ **Leica I**, launched in 1925.

# Visual tour

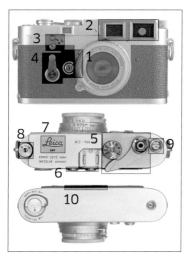

**KEY**

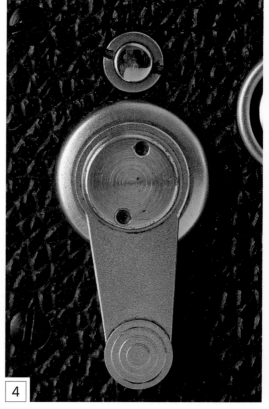

➤ **LENS AND MOUNT** Leitz made a range of lenses for the M3, which were attached to the body with a new bayonet mount rather than the screw mount used on earlier models. This made changing lenses both quicker and easier. Fitting a lens was simply a matter of aligning two red dots and twisting the lens anticlockwise for about one-eighth of a turn. It was even possible to change a lens with just one hand.

▲ **VIEWFINDER WINDOWS** The rangefinder focusing aid works by forming two images in the viewfinder. As the focusing control is moved, the two images come together, and when the images coincide exactly, the lens is correctly focused.

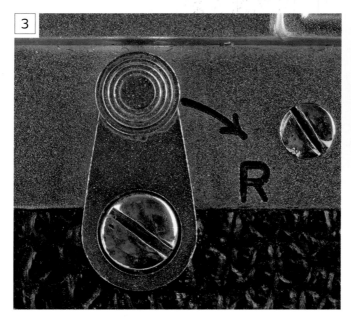

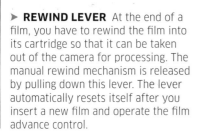

➤ **REWIND LEVER** At the end of a film, you have to rewind the film into its cartridge so that it can be taken out of the camera for processing. The manual rewind mechanism is released by pulling down this lever. The lever automatically resets itself after you insert a new film and operate the film advance control.

▲ **SELF-TIMER** The M3 was the first Leica to have a self-timer, operated by a simple manual lever on the front of the camera. This made it possible to delay the shutter's operation for five to ten seconds, so that photographers could include themselves in photographs.

▶ **FILM ADVANCE LEVER** Pulling this lever twice moves the film on by one frame and cocks the shutter. Leica believed that two short strokes were better than a single long stroke because users would not have to alter their grip on the camera, making rapid shooting easier.

5

▶ **FLASH TERMINALS** On the back of the camera are two terminals, one for using a flash gun with bulbs and the other for an electronic flash. When it is connected, the flash fires as the camera's shutter opens. Electronic flash works at speeds up to 1/50 sec and flash bulbs at up to 1/1000 sec.

6

## ON **DESIGN**

Leica cameras were first produced when most good quality photographic equipment was large, cumbersome, and often designed for studio use. Early cameras had wooden cases, had to be focused using leather bellows, and stood on big tripods. The Leica was different. It was designed to be a high-quality camera that could go anywhere. The metal body was robust, the dials beautifully machined, the range of features impressive, and the whole camera very compact. Leica added one item that was not built in, but was designed to be used in conjunction with its cameras – an exposure meter. This measured the light level and displayed suggested shutter and lens aperture settings, eliminating much of the guesswork from photography. Like the camera, the meter was housed in a strong metal body. It was designed to sit on top of the camera, but you could hold it separately and operate it with one hand. This practical piece of equipment was another fine example of what Leitz's fusion of design and engineering could produce.

▲ **Detachable exposure meter** for the M3 Rangefinder

◀ **LEICA NAMEPLATE** The camera's top plate is engraved with the names of both Leica and Ernst Leitz, the former in the company's trademark cursive script. By the time the M3 was launched, this name had become synonymous with quality and the company's products were sought out by both professional photographers and keen amateurs.

7

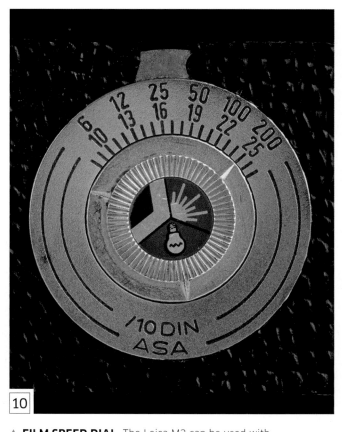

8

9

10

▲ **SHUTTER SPEED DIAL** This dial on the camera's top plate allows you to set the Leica's beautifully engineered shutter to speeds between 1 and 1/1000 sec. This was a wide range for the time, with the fast 1/1000 speed allowing the camera to take good pictures with sensitive film in bright sunlight, and also to freeze subjects in motion.

▲ **FRAME COUNTER** Positioned just to the right of the shutter release and film advance lever, this numerical read-out shows how many frames of the current film have been used. The dial counts up to 40 exposures and resets itself when you load another cartridge of film.

▲ **FILM SPEED DIAL** The Leica M3 can be used with various types of film, which vary according to their speed (sensitivity to the light). On the back of the camera is this indicator, which enables you to set a pointer to the speed of the film you are using, as a reminder. The indicator displays both German DIN ratings and the equivalent American ASA ratings, so that you can load film rated using either system.

# Butterfly stool

1954 ■ FURNITURE ■ PLYWOOD AND STEEL ■ JAPAN

SCALE

## SORI YANAGI

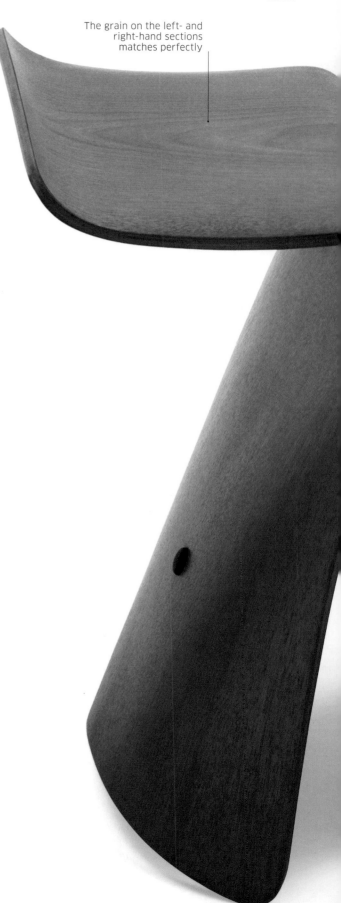

The grain on the left- and right-hand sections matches perfectly

**Modern ideas and materials** combined with traditional values and craftsmanship can result in great design. A fascinating example is the Butterfly stool, created by Japanese designer Sori Yanagi in 1954. The stool is strikingly simple, consisting of two identical pieces of bent plywood, joined where they meet and reinforced with a single metal rod. Yanagi chose a modern material – plywood – because it is strong, attractive, and can be bent and moulded under pressure to form interesting shapes, as Charles and Ray Eames found when they developed their celebrated Lounge Chair (see pp.152–55). The form Yanagi created was also modern in its purity and economy.

The stool is beautifully crafted – Yanagi used only the finest materials, so that the plywood featured the patterned grain of high-quality timbers such as rosewood or maple. The elegant shape of the stool has been compared both to the strokes used in Japanese calligraphy and to *torii*, the gateways that form the approaches to Shinto temples. Even though the stool is not a typical piece of furniture in traditional Japanese houses, Yanagi managed to make this one seem part of the culture of his homeland.

This fusion of ancient and modern reflects Yanagi's upbringing and training. His father, Soetsu Yanagi, was founder of the Japanese Folk Crafts Museum in Tokyo and a leader of the Mingei movement, which advocated using traditional crafts as the basis of new work and championed the work of artist-craftsmen. However, Sori Yanagi was also influenced by Western Modernism and worked with French designer Charlotte Perriand (see p.56) when she lived in Tokyo in the 1940s. Much of Yanagi's work is a perfect synthesis of Eastern and Western traditions.

---

### SORI **YANAGI**

#### 1915–2011

Sori Yanagi trained as a fine artist in Tokyo, but became interested in the ideas of Le Corbusier (see p.57) and went to work in an architectural studio. He met Charlotte Perriand through his employers, and worked as her assistant in the 1940s before setting up his own industrial design studio. In the 1950s, he played a major part in promoting industrial design in Japan, both through his own practice and as co-founder of the Japan Industrial Designers' Association. His work spanned many different areas of design, from highways and bridges to glassware and furniture, including the famous stacking or "elephant" stool. Yanagi was also commissioned to created the torches for the 1964 Olympic Games, which were held in Tokyo.

# "Things that are easy to use survive..."

**SORI YANAGI**

# Visual tour

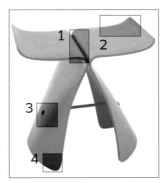

**KEY**

➤ **CENTRE LINE** The two sheets of plywood come together at the top to make a distinctive groove, marking the centre of the seat. From above, it is hard to see at first how the stool is held together, but the fixing bolts are just visible beneath the bends of the seat.

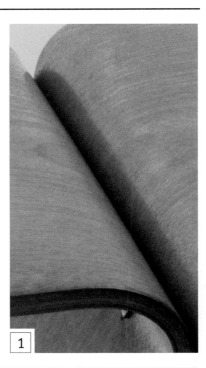

1

2

▲ **WINGS** The seat flicks gently upwards at the edges, reminiscent of a butterfly's wings, which gives the stool an organic feel and helps to frame the seat.

➤ **STRETCHER ROD** The stool is strengthened by the metal rod that links the two curved legs, similar to a stretcher on a traditional chair. The rod's fixings are finished with a simple metal disc as they pass through the plywood.

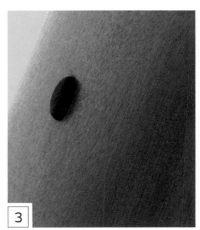

3

The stretcher rod finishes vary – some are in stainless steel, others have a bronzed finish

The edge of the leg curves slightly above the floor, so that the stool seems to float

◄ **FEET** The plywood is shaped and bent so that it touches the floor at each corner like four feet, giving the stool stability and absorbing the weight of the sitter. The edges of the feet are gently curved – there are no hard angles anywhere on the stool.

4

■ **The Man with the Golden Arm poster** Saul Bass

■ **Citroën DS** Flaminio Bertoni

■ **Apple vase** Ingeborg Lundin

■ **Phonosuper SK4** Dieter Rams, Hans Gugelot

■ **Mirella sewing machine** Marcello Nizzoli

■ **Tulip table** Eero Saarinen

■ **Lounge Chair 670 and ottoman** Charles and Ray Eames

■ **Wall clock** Max Bill

■ **Helvetica typeface** Max Miedinger

■ **Egg chair** Arne Jacobsen

■ **PH Artichoke light** Poul Henningsen

■ **Mono-a flatware** Peter Raacke

■ **Austin Seven Mini** Alec Issigonis

■ **Cadillac series 62** Harley Earl

■ **Panton chair** Verner Panton

1955–1966

# The Man with the Golden Arm poster

1955 ▪ GRAPHICS ▪ COLOUR PRINT ON PAPER ▪ USA

SCALE

## SAUL BASS

**Saul Bass was one of the most renowned** graphic designers of the 20th century. Much of his work was in film, where he liked to design a whole visual identity for a movie – publicity material, the cover for the original soundtrack recording, and the title sequence, as well as the poster. Among his best-known designs is the poster artwork for Otto Preminger's *The Man with the Golden Arm*, a movie in which Frank Sinatra plays a poker dealer and would-be jazz musician, who is addicted to heroin. The main image on the poster is not a photograph of Sinatra's face, as would have been normal at the time, but the jagged silhouette of an arm, symbolizing the character's battle with drug addiction. This dramatic image, which slices through the film's title in a graphic gesture of violence, is set against the informal lettering of the title and credits, and the broad blocks of colour. The portraits of the film's stars were added later at the insistence of the film studio. Preminger was so impressed with the image that he asked Bass, who was interested in animation, to integrate the arm into the movie's opening title sequence. This began with a series of abstract black bars that eventually merge to form the arm and fingers. Both the titles and poster combine great graphic sophistication with a disturbing quality that reflects the film's disquieting subject matter.

### SAUL **BASS**

#### 1920-96

New Yorker Saul Bass studied commercial art before taking a course in Bauhaus-influenced advertising design under Gyorgy Kepes, a painter who wrote and lectured on visual culture and technology. After moving to California, Bass ran a successful studio, designing logos and corporate identities for major companies such as AT&T, Warner, and Continental Airlines. His big break in films came from Otto Preminger, who hired Bass to work on *Carmen Jones* in 1954. A long string of film commissions followed, including the animated title sequence for *Around the World in Eighty Days* and many collaborations with Alfred Hitchcock. Later, Bass worked for Stanley Kubrick (*Spartacus* and *The Shining*), Steven Spielberg (*Schindler's List*), and Martin Scorsese (*Goodfellas*).

# Visual tour

KEY

▶ **THE HAND** Jagged, like the arm above it, and claw-like, the hand has simple, elongated wedges of black to represent the fingers. It is positioned slightly off-centre and stands out dramatically against the stark white background, appearing to clutch at empty space, in a gesture of despair.

1

2

▲ **LETTERING** Saul Bass worked directly with master calligrapher Maury Nemoy to create the distinctive lettering. The characters are irregular in height, thickness, and position, creating an unsettling effect.

▶ **BLOCKS OF COLOUR** These broad areas of colour, like torn strips of paper, extend across the poster. They have uneven edges and are set at slightly odd angles – another means by which Bass adds to the sense that nothing is quite as it should be.

3

# Visual tour

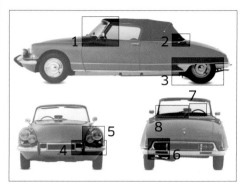

**KEY**

▶ **WINDSCREEN** During the 1950s, wraparound windscreens became increasingly common, and the one fitted to the DS has a gentle aerodynamic curve. This wraparound screen not only made the car look good, but helped improve the driver's view by minimizing blind spots caused by the supporting pillars.

▼ **REAR INDICATOR** On the hardtop version of the DS, the rear indicator was positioned on the rear pillar, where it was highly visible. As the convertible has no rear pillar, the indicator is placed slightly lower down. The housing curves around the corner, so that it can be seen both from behind and from the side – a valuable safety feature.

▲ **REAR WHEELS** The treatment of the wheels on the DS is very distinctive. Full-diameter hubcaps cover everything except the tyre. The rear wheel is partly concealed behind the bodywork, which cuts across the upper part of the wheel in a straight line. This gives the rear of the car an aerodynamic appearance and was adopted on several later Citroëns.

◀ **FRONT** The lack of a conventional radiator grille is one of the details that makes the front of the DS particularly eye-catching. Instead, there are small openings below the bumper to let in air to help cool the engine. Other imaginative design features include the wraparound indicator lights.

▼ **FRONT HEADLAMPS** The original DS had a single pair of headlamps but this 1963 convertible has auxiliary lamps, and both sets of lamps have chrome rims. These are set in a fixed position, but on some models, the smaller lamps swivelled as the car turned a corner.

▼ **REAR LIGHTS** Set on either side of the number plate, the rear lights and brake lights are recessed slightly, to make the back of the car look tidy. These very simple fittings, surrounded by the rear chrome trim, are scarcely noticeable until they are turned on and the neat red circles light up.

5

6

7

▲ **STEERING WHEEL AND DASHBOARD** The DS has a striking single-spoke steering wheel, which surprised people when they first saw the car. Its practical purpose was to provide the best possible view of the array of dials on the dashboard.

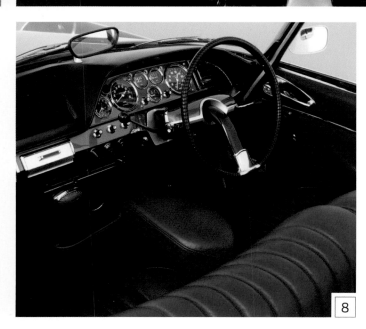
8

◀ **FRONT SEATS** The DS was always a luxury car and Citroën emphasized passenger comfort. The cars at the top end of the range had deep, leather-upholstered seats and a padded arm rest between the driver and the front passenger. Everyone on board was cradled in luxury as they enjoyed the vehicle's smooth, even ride.

## ON **DESIGN**

The DS was one of a series of ground-breaking car designs produced by Citroën. The 7CV and Traction Avant range of front-wheel drive cars with integrated chassis and body were the first, dominating Citroën's range in the 1930s and 1940s. Then came the tiny 2CV, also known as "the duck", styled by Bertoni and introduced in 1948. This was the most basic of cars, but it was solidly built, could be driven on rough ground, and its twin-cylinder engine was both reliable and economical. To everyone's surprise it sold in huge numbers. With both the DS and 2CV in production after 1955. Citroën proved that it could produce well-designed vehicles for a range of markets.

▲ **Citroën 2CVs** in the factory, 1953

## IN **CONTEXT**

When the DS was launched at the Paris Motor Show on 5 October, 1955, the response was remarkable. The car was widely admired and, after just one day, the company had taken some 12,000 orders. This enthusiasm was reflected by the critic Roland Barthes, who said that the car looked as if it had "dropped from the sky". He was also delighted by the car's initials: in French, the letters "DS" are pronounced exactly the same as the word for goddess, *déesse*.

The launch came at a difficult time for France, which was still recovering from World War II as well as facing the decline of its colonial power. This stunning example of French design and technology gave the nation hope.

▲ **Crowds admire the DS** at its Paris launch

# Apple vase

1955 ▪ GLASSWARE ▪ BLOWN GLASS ▪ SWEDEN

## INGEBORG LUNDIN

SCALE

**In many of the most innovative glass designs** of the 20th century, colours or patterns have been incorporated to add interest, or the form of the glass has been manipulated into unusual shapes. The Apple vase, designed by Ingeborg Lundin for the Swedish glass company Orrefors, works in the opposite way. Lundin has reduced blown glass to its basic elements – purity of form, translucency, and simplicity. Her idea was to design a bulbous vessel in the shape of an apple – a swelling form for the body and a short neck that represents the fruit's stem. The vase was created using a technique called flashed glass, in which a layer of coloured glass lies beneath one of clear glass, giving the colour a pale, slightly unearthly quality. Producing this kind of glass with the exact degree of precision required is a highly skilled task and Orrefors selected their most accomplished glassblowers to make it. Lundin also designed a smaller, slightly flatter version of the Apple, the Melon, which demanded similar accuracy.

The Apple vase, which was originally produced for the Helsingborg 55 exhibition, was also entered in the Milan Triennale in 1957. After winning a gold medal, the vase soon became a commercial success: people responded to the combination of the whimsical apple shape and the subtle colour of the glass. This exquisite object, which is not so much a vase as a piece of sculpture in glass, seemed to encapsulate the quality of purity characteristic of so much Swedish design of the 1950s, and its popularity helped to boost Sweden's reputation as a country that was producing some of the best in design for the home.

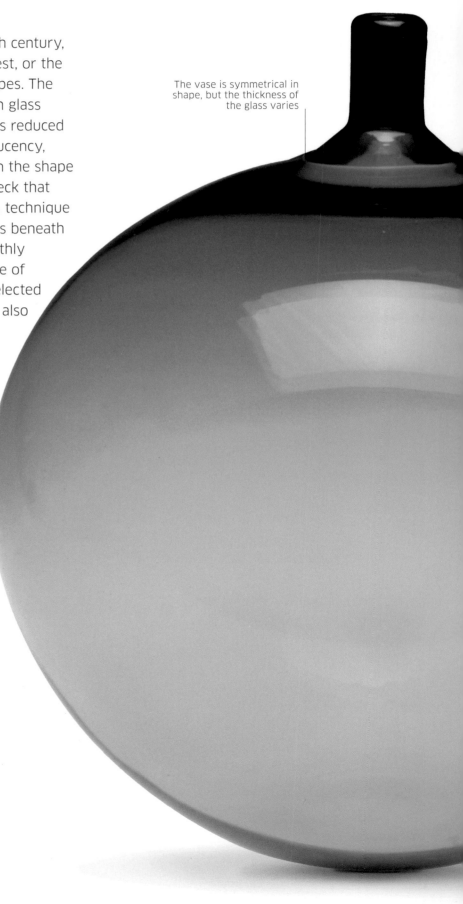

The vase is symmetrical in shape, but the thickness of the glass varies

### INGEBORG **LUNDIN**

#### 1921-92

Swedish artist Ingeborg Lundin trained in Stockholm. In 1947, she was the first female designer to be taken on by the celebrated glass manufacturer, Orrefors. She remained with the company for some 24 years, producing many designs that exploited the creative skills of the glassblowers, and her work was widely exhibited. The publicity provided by exhibitions such as Design in Scandinavia, which toured the USA in the 1950s, put her at the forefront of Swedish design. Lundin won the Lunning Prize, awarded to distinguished Scandinavian designers, in 1954.

# Visual tour

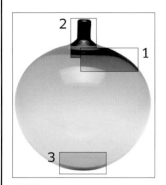

**KEY**

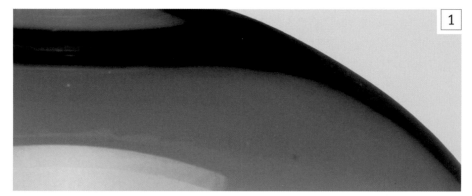

1

◄ **BODY** The glass of the body of the vase is slightly uneven in thickness, creating variations in the intensity of the green. Dark and opaque at the top and bottom, the glass is almost transparent in the middle of the vase. The colour also varies when the light conditions around it change.

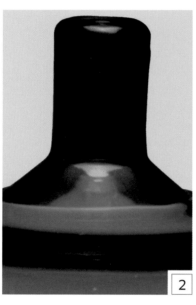

2

3

◄ **NECK** The short neck is narrow and slightly flared towards the top, just like an apple's stem. Although it is possible to pour water into the vase and use it as such, this piece of art glass was designed to be a sculptural object in its own right.

▲ **BASE** The base of the vase varies in thickness, and this catches the light. The base is also slightly flattened, to make it look like the bottom of an apple, and this gives the vase enough stability to stand firm on a surface.

## ON **DESIGN**

Among Ingeborg Lundin's other successful designs were vessels made using a highly complex process that was invented at the Orrefors glassworks, in the Swedish province of Småland. After creating a piece of blown glass with two layers of different colours, areas of the outer layer were masked and the vessel sandblasted to create a pattern in two colours. The glassworker then applied a third layer of glass, which trapped air pockets that had been created in the sandblasted areas to give the vessel its distinctive appearance. Edward Hald of Orrefors named this technique Ariel, after the spirit of the air in

Shakespeare's play *The Tempest*. Lundin used it to create several unusual pieces with both abstract and figurative designs. The abstract patterns in browns, blues, and greens are made up of stripes or simple shapes such as squares or discs, which stand out against a clear background. The most common figurative examples show profiles of human heads depicted in clear lines against a coloured background. All these vessels have thick walls because of their multilayered construction, but although this makes them feel heavy, the decoration creates the impression of light and air.

The vase was produced in sizes varying from 30cm (12in) to 50cm (20in) in diameter

▲ **Ariel vase No. 534**

▲ **Profiles vase**

# Phonosuper SK4

1956 ■ PRODUCT DESIGN ■ PAINTED METAL, WOOD, AND ACRYLIC PLASTIC ■ GERMANY

SCALE

## DIETER RAMS, HANS GUGELOT

**In 1955, the influential German designer** Dieter Rams joined the Braun company. His first product, designed with Hans Gugelot of the Ulm School of Design (see p.152), was the Phonosuper SK4 – an innovative new form of radiogram. In contrast to earlier models, which usually had wooden cases made by cabinet-makers, the SK4 radio and record player were housed in a sleek, modern box of steel and wood with no ornamentation. The controls were simple, plastic knobs, and the loudspeaker grille a series of horizontal slots in the front. These features made the SK4 a pure example of the Ulm School's systematic approach to design – an object had to be clearly thought out and easy to use, with no extraneous flourishes. The position of the controls, for example, was entirely logical: they were all together on the top of the unit, rather than on the top and front, as on many earlier radiograms.

The resulting clean, modern-looking design was instantly popular and fitted well into contemporary homes. It was followed by many other successful Braun products that brought a similar rational and minimalist aesthetic to the market, helping to raise standards of design in consumer electronics across Europe and beyond.

## "Good design is as little design as possible"

**DIETER RAMS**

Projecting lugs support the record on the turntable

Slots are cut in the casing in front of the speaker

### DIETER **RAMS**

#### 1932-

After training as an architect and designer in Wiesbaden, Dieter Rams worked as an architect. When he joined Braun in 1955, he started by designing exhibition sets and offices, then became more interested in products. Head of the design department from 1961 to 1995, his work encompassed a wide range of products, from alarm clocks to the ET44 pocket calculator (with Dietrich Lubs), but he became best known for his audio equipment and television sets, including the T1000 world receiver and the Studio 2 system. From 1959, Rams also designed seating and storage for the furniture company Vitsœ.

## ON **DESIGN**

Dieter Rams firmly believed that a designer should strive to make each product innovative, useful, easy to understand, and durable. Every detail needed to be carefully thought out and functional. Rams also thought that designs should be attractive, but in an unobtrusive way. This approach meant that groups of items designed for Braun have a strong family resemblance. The electronic products that Rams designed in the 1950s and 60s, for example, often had metal cases, slots or holes for loudspeaker grilles, and simple geometric controls. His later designs, such as the clocks and calculators he produced in the 1970s, had black plastic casings, white numerals, and clear, sans serif type, creating a similar effect of unity and purposefulness.

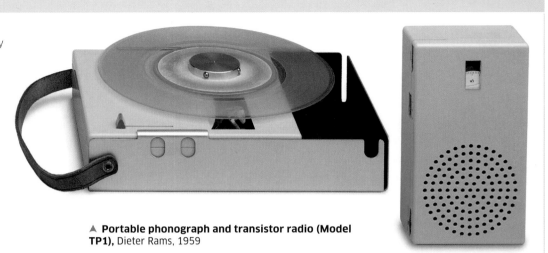

▲ **Portable phonograph and transistor radio (Model TP1),** Dieter Rams, 1959

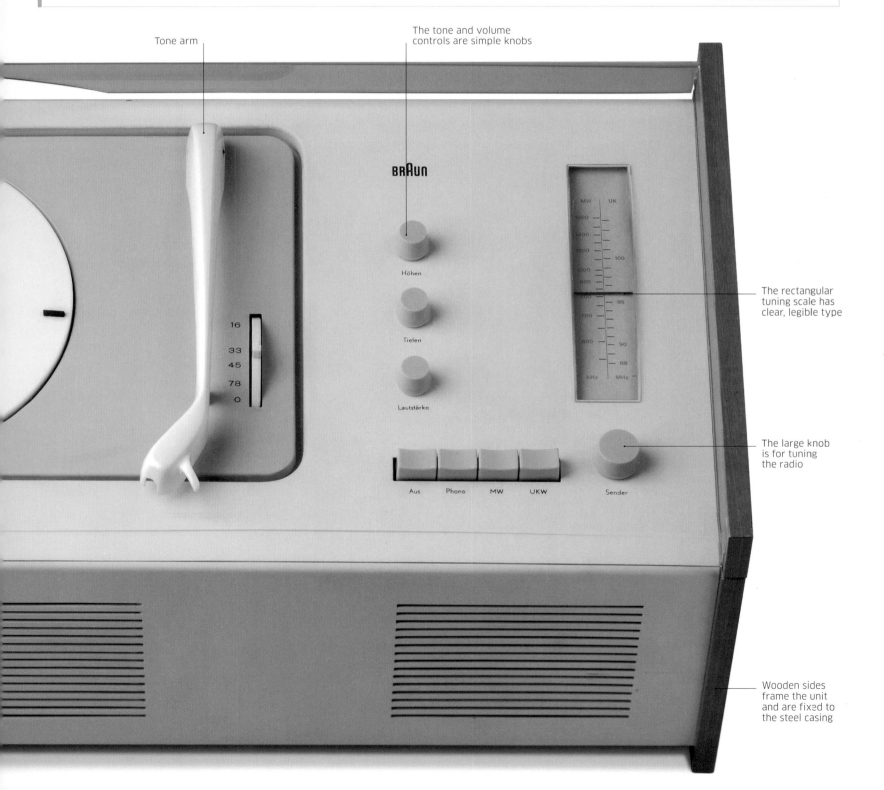

Tone arm

The tone and volume controls are simple knobs

The rectangular tuning scale has clear, legible type

The large knob is for tuning the radio

Wooden sides frame the unit and are fixed to the steel casing

# Visual tour

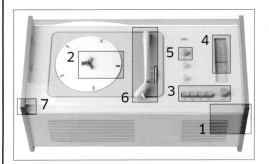

**KEY**

▶ **CABINET** The front, top, and rear of the cabinet are formed from a single piece of sheet steel that wraps around the unit, making a gentle curve at the front edge and butting up to the wooden sides. The resulting lack of joints gives the SK4 a clean, seamless appearance and enabled the manufacturers to insert the working components through the base.

1

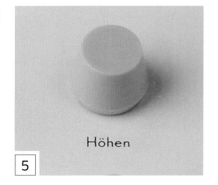

5

▲ **CONTROL KNOBS** Three knobs control the sound produced by the SK4 - one for tone, one for volume, and one for balance. Unlike many similar knobs, they are not surrounded by a numerical scale. Users were expected to adjust the sound by ear to achieve their preferred balance.

2

◀ **TUNING DIAL** The SK4 was designed to receive both AM and FM broadcasts, which were tuned using this analogue dial. The dial is straight rather than round, in contrast to the dials of many earlier radios. It has a red bar and clear, sans serif type that makes it easy to read.

▲ **TURNTABLE** Unlike many turntables of the time, which had a raised surface and a dark rubber mat to protect the fragile surfaces of records, the SK4's is finished in white and is almost flush with the cabinet. The protrusions around the edge support the records and raise them slightly, so they are easy to lift on and off the turntable.

4

▶ **BUTTONS** A row of push buttons enables users to select a radio band or the record player. These buttons look completely different from the circular knobs for the volume and tone because they have a different purpose - this is a good example of how the appearance of the SK4's features is linked to their function.

3

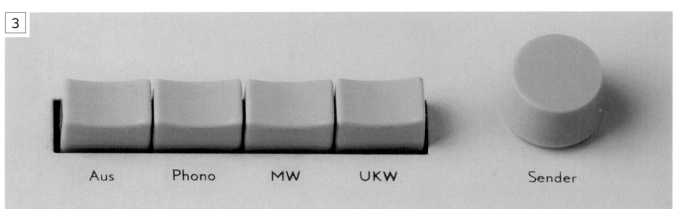

◄ **TONE ARM** The tone arm is the one feature of the SK4 that deviates slightly from its otherwise pure geometry. The arm tapers in shape and there is a small protrusion for the finger, designed to make the arm easy to pick up and lower onto records. The speed of the turntable, which can be adjusted for the four different types of record in use at the time, is selected using a control to the right of the arm.

▼ **LID** The design is completed by a transparent lid. This was originally going to be made of the same steel as the main part of the SK4, but the prototype affected the unit's sound quality, so Rams chose Plexiglas – the first time this material had been used on a mass-market consumer product. The clear lid and the pale colour of the SK4 earned it the nickname "Snow White's Coffin".

## ON **DESIGN**

Dieter Rams and Hans Gugelot began a design by breaking an item down into its most basic elements and analysing both its function and geometry. Early sketches for the SK4 show little more than a series of circles and rectangles, but the basic form of the unit is already beginning to emerge. This approach not only kept the design clear and minimal, it also helped the designers to produce items with simple shapes that would not be difficult to manufacture. German designers in the 1950s paid more attention to the demands of industrial production methods than the more craft-oriented Modernist designers had done in the 1930s.

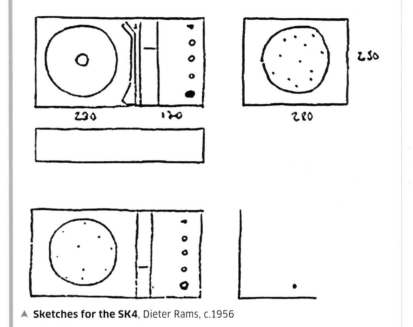

▲ **Sketches for the SK4**, Dieter Rams, c.1956

## IN **CONTEXT**

In the 1950s, German designers focused on geometrical simplicity, eliminating unnecessary detail and ornament, and on technology. This approach was influenced by the ideas of designers such as Hans Gugelot at the Ulm School of Design, who began to work as a consultant for Braun at around the same time that Dieter Rams joined the company. Their way of working fostered a rational attitude to design that became very influential – notably in consumer electronics and appliances, but also in other areas. It even affected the design of children's toys. When the industrial and furniture designer Walter Papst created his Rocking Sculpture, the addition of a tail was sufficient to let people know that the red plastic toy was a stripped-down version of the traditional rocking horse.

▲ **Rocking Sculpture**, Walter Papst, 1959

# Mirella sewing machine

1956 ■ PRODUCT DESIGN ■ ENAMELLED ALUMINIUM AND OTHER METALS ■ ITALY

SCALE

## MARCELLO NIZZOLI

**During the 1950s, when German designers** such as Dieter Rams (see p.144) were revolutionizing product design with their uncluttered, functional approach, manufacturers in Italy were moving in a different direction. Italian products were also functional, but tended to be more sculptural in appearance, favouring organic curves rather than the disciplined geometry of German designs. The streamlined appearance of Marcello Nizzoli's Mirella sewing machine, designed for the Necchi company in 1956, is a striking example.

Necchi's brief was to design an electric sewing machine that was easy to use and attractive to customers who had previously had manual machines with hand cranks or treadles. Nizzoli gave the machine a detachable crank so that it could also be operated by hand, a simple straight-stitch mechanism, and a curvaceous exterior with few hard edges. In addition, the Mirella had a light that shone on the needle mechanism when the machine was in use – a welcome innovation for anyone used to a hand-operated model. Marketed as a machine that was sophisticated yet simple to operate, the design won praise from both users and the design community. The Mirella was awarded the 1957 Compasso d'Oro, confirming the success of this sculptural, organic approach in Italian design, and it remained in production until the early 1970s.

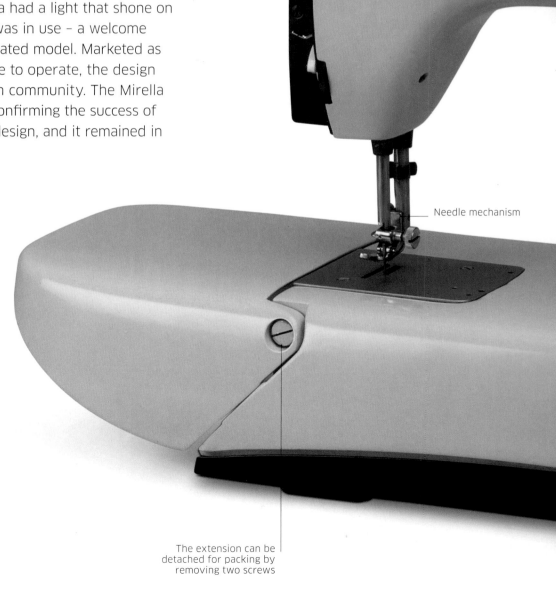

The curvaceous shape became closely identified with Necchi machines

Needle mechanism

The extension can be detached for packing by removing two screws

### MARCELLO **NIZZOLI**

**1887–1969**

Educated at the Accademia delle Belle Arti in Parma, Italian artist, architect, and designer Marcello Nizzoli began work as a draughtsman in Milan. After World War I, he diversified, taking on projects that included advertising posters and fashion accessories, and absorbing the influences of art movements such as Futurism and Cubism. In the 1930s, he became part of the Italian Rationalist group of architects, working on various buildings in Milan, and began a fruitful association with the Olivetti company. Nizzoli's work for Olivetti included the Lettera 22 typewriter (see p.195), which displayed all the hallmarks of his approach to design – good ergonomics, clearly identified keys and controls, and an attractive, curvaceous exterior. Other products ranged from cigarette lighters to a petrol pump.

## ON **DESIGN**

Marcello Nizzoli was a talented graphic designer. His clients included Campari, which ran high-profile poster campaigns in the interwar period, as well as other Italian companies, such as Fiat. Nizzoli produced striking posters to promote tourism. Although some of his work featured realistic images of items such as cars or drinks bottles, he also produced more abstract designs, such as the Milan poster (right), in which the stylized pinnacles and blocks of rich colour represent the city's cathedral and its stained glass. All of his graphic work, whether abstract or realistic, was very clear and easy to read.

▶ **Milan Tourism poster**, Marcello Nizzoli, 1955

*mirella*

The wheel can be turned by hand to raise the needle

# Visual tour

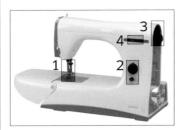

## KEY

▶ **NEEDLE MECHANISM**
The Mirella has a generous area around the needle and needle plate, to help the user control the fabric. A guide can be attached to the needle plate, to make it easier to sew seams parallel to the edge of the fabric. The screw on the needle mechanism can be loosened or tightened to remove or insert a needle.

▶ **CONTROLS** The main control for regulating the stitch length is on the front of the machine. It is marked with clear numerals in white for the different stitch lengths. The smaller button marked "R" switches the machine to sew in reverse.

1

2

3

◀ **REAR PLATE** The dark side plate is recessed slightly from the lighter front and back of the machine, emphasizing the way the enamelled aluminium casing curves around it. The side plate contains the hand wheel and the socket for the crank that makes it possible to operate the machine by hand.

▼ **NAMEPLATE** The Necchi name is clearly displayed in simple sans serif letters on a dark plate. An established sewing machine manufacturer from the early 20th century, Necchi had a strong brand identity that it confirmed with prominent typography.

4

# Tulip table

1956 ■ FURNITURE ■ CAST ALUMINIUM BASE AND WOOD, LAMINATE, OR MARBLE TOP ■ USA

SCALE

## EERO SAARINEN

**Several leading furniture designers** in the 1950s sought to combine the uncluttered aesthetic of Modernism with a sculptural quality, using new materials to create strong shapes. One centre of this movement was Cranbrook Academy of Art in Michigan, where designers such as Harry Bertoia (see p.120) were carrying out innovative work. The academy had been co-founded by architect Eliel Saarinen, a Finnish immigrant. His son, Eero, who studied at Cranbrook, cultivated an expressive quality in his furniture designs, aiming for a more organic form of Modernism. He became famous for the Tulip range, its elegant tables and chairs supported on a central round pedestal. Saarinen said he wanted to improve on the cluttered effect of conventional tables and "clear up the slum of legs" characteristic of most furniture. He designed a single, trumpet-shaped pedestal that gradually tapers from a broad top to a slender, stem-like upright before swelling out again until it forms an almost flat disc at the base. The designer initially specified cast iron for the pedestal, but later used aluminium, covering it in a white, plastic coating. The metal base was heavy enough to support a large table top and, although the original tops were plain white like the support, variations were later produced, including marble tops coated with a clear, stain-resistant plastic. The manufacturer, Knoll, produced both full-size dining tables and occasional tables (as shown), with either round or oval tops. The Tulip table was widely admired because although its seamless, sculptural shape made a strong statement, it also went well with more traditional furnishings. When combined with Saarinen's Tulip chairs, however, the table looked almost futuristic. It quickly became a design classic.

## "Always design a thing by considering it in its next larger context"

**EERO SAARINEN**

### EERO **SAARINEN**

#### 1910-61

Finnish-American architect Eero Saarinen enrolled at Cranbrook Academy, Michigan, where he met fellow students Charles and Ray Eames (see p.152), before studying sculpture in Paris and architecture at Yale. He became famous as a furniture designer in the 1940s, winning an award with Charles Eames, and creating successful designs such as the Tulip range and the Womb chair, both produced by Knoll. As an architect, his buildings display a variety of styles, from the curvaceous St Louis Gateway Arch and the swooping TWA Terminal building, New York, to rectilinear structures in the Modernist idiom.

### IN **CONTEXT**

Eero Saarinen developed an interest in moulded furniture in the early 1940s, when working with Charles Eames on their winning entry for the competition "Organic Design in Home Furnishing", organized by New York's Museum of Modern Art. Eames' and Saarinen's seating was in moulded plywood, but by the 1950s, Saarinen found that fibreglass was perfect for forming sculptural chairs that followed the contours of the human body. When he was working on the Tulip table, he developed the shell of the Tulip chair – named for its resemblance to the flower – out of moulded fibreglass. The material was not strong enough for the pedestal base, which, like his tables, was made from aluminium but covered in white to create uniformity.

► **Tulip chair**, Eero Saarinen, 1956

The original table tops were made of coated wood

The top tapers gently towards the edge to reflect the curves of the base

The base is made from an aluminium casting that has been polished and given a protective white coating

# Visual tour

**KEY**

1

2

▶ **BASE** On traditional wooden pedestal tables, the feet have to extend almost as far as the table top for the structure to be stable. The base of the Tulip table is very heavy, so it does not need to be so wide. The central support terminates in a circular pedestal base about a third as wide as the table top.

1

2

▶ **TABLE TOP** Both the table top and pedestal base were produced in a white finish, to create uniformity and emphasize the sculptural quality of the table. The play of light on the table top created variations of tone and in most settings, the brightly lit top would contrast with the shadows cast on the central support.

# Lounge Chair 670 and ottoman

1956 ▪ FURNITURE ▪ LAMINATED WOOD, ALUMINIUM, STEEL, AND LEATHER ▪ USA

SCALE

## CHARLES AND RAY EAMES

**The most luxurious chair** in the mid-century Modern style was created by American designers Charles and Ray Eames, and known simply as Lounge Chair 670. The chair and its accompanying ottoman marry rich materials (leather upholstery, laminated wood veneered in rosewood, and immaculately finished aluminium) with a purposeful design in which the chair's main elements are clearly defined.

Originally designed as a present for the film-maker Billy Wilder, a friend of Charles and Ray, the chair was a result of the Eames's early attempts to bend plywood using heat and pressure in 1940. In the same year, furniture by Charles Eames and fellow designer Eero Saarinen (see p.150) won first prize in a competition sponsored by The Museum of Modern Art in New York, spurring them on to more plywood designs. The 670 was the most lavish and one of the most famous of these. Charles Eames said that the chair's shape was inspired by the appearance of a "well-used first baseman's mitt", and its generous size and plump cushions embrace the sitter as effectively as an old leather club chair. However, the 670 is also undeniably modern and furniture manufacturer Herman Miller (see p.109) was happy to market it alongside other icons of modern design. The luxurious chair continues to be a surprisingly good seller.

### CHARLES AND RAY **EAMES**

#### **1907-78** AND **1912-88**

Charles Eames trained as an architect and worked for a firm in his native St Louis, Missouri, USA, before setting up his own practice in the town. In 1936, he was awarded a fellowship at the Cranbrook Academy of Art, where he met both fellow designer Eero Saarinen and his future wife, Ray Kaiser, an abstract artist. Working with Ray, Eames designed a huge range of furniture, as well as creating exhibition designs and architectural projects and making over 100 short films. Many of these designs were originally attributed solely to Charles, but the close, collaborative nature of the couple's work and the extent of Ray's contribution is now recognized.

The cushions have a plastic backing that is fixed to the wood with concealed clips and rings

The black base of the chair matches the leather upholstery

Buttons emphasize the depth and luxury of the cushions

Richly figured rosewood veneer

Angled back for comfort

▲ **Front view**

▲ **Rear view**

The shape of the cushion follows the contours of the wooden shell

Five-ply timber was originally used for the shells; modern chairs and ottomans are usually made from seven-ply

The chair base has five supports to make it completely stable

"Whoever said that pleasure wasn't functional?"

**CHARLES EAMES**

# Visual tour

**KEY**

> **ARMREST** The comfortably padded armrest is the component that joins the back and seat. The metal frame of the arm is attached to the wooden back and seat shells by means of a "shock joint" with a thick rubber washer. This allows the chair to flex slightly, adding to the feeling of gentle relaxation and comfort as you sink into the cushions and the chair moulds itself to your body.

1

2

▲ **BACK CUSHION** The cushion fits the back precisely and its piped edge enhances the feeling of luxury. The cushions were originally stuffed with goose feathers, but artificial foam filling is used for the more recent models.

> **REAR SUPPORTS** A striking feature of the chair is the way in which the back is in two sections – one supporting the sitter's lower back and the other the shoulders. The sections are linked by a pair of exposed cast aluminium bars bolted to the shells with rubber discs. The bolts can be removed and the whole chair dismantled for transportation.

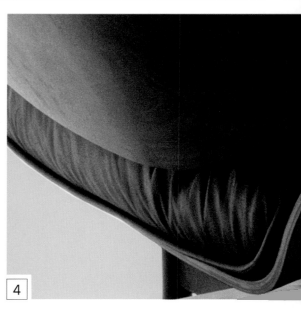

3

4

▲ **REAR OF SEAT AND BACK** The seat and back of the 670 seem to float independently of each other, with a small gap where the two leather-covered cushions are visible. This clever effect is possible because the armrests act as the links between the two components, giving the chair its distinctive appearance. As in many of the Eames's designs, there is a clear separation between the supporting and the supported parts.

**◄ PLYWOOD SHELL** ⑤

The curved profile of the chair's plywood shell was first developed by Eames in the 1940s, using his own special press. By the time he and Ray designed the 670 chair, it was possible to create a more dramatic curve. This was an ideal way to show off the rich grain of the rosewood veneer, which forms a contrast with the yellowish inner layers of the plywood.

**◄ SEAT CONNECTION** ⑥

The connection between the seat and the base of the chair is critical to the overall comfort that it provides. Not only does it enable the chair to achieve a tilted rocking motion, but it allows it to swivel through 360°.

**◄ CHAIR FOOT** ⑦ The glides at the end of each of the five base supports are adjustable to compensate for any unevenness in the floor. This practical feature is now standard on all kinds of furniture but was unusual when the 670 was first manufactured.

## ON **ARCHITECTURE**

Most of the buildings designed by Charles and Ray Eames were houses. The most celebrated were built as part of the Case Study Program, a project sponsored by *Arts & Architecture* magazine with the aim of producing inexpensive, well-designed homes using modern materials. The Eames's own house was part of this project. Its steel frame, with sliding windows and walls, was uncomplicated and quick to assemble. The primary-coloured walls reminded more than one visitor of paintings by Mondrian, and the light, airy, interiors with well-chosen decorative objects were greatly admired. What particularly impressed critics about this one-off, carefully designed house was the fact that many of the components, including the windows and doors, were standard items from manufacturers' stock. Charles and Ray Eames designed other houses, including one for Billy Wilder that was never built, but none of their architectural work was as successful or influential as the house that they built for themselves.

▲ **The Eames House**, Case Study House No. 8, Santa Monica, California, USA, 1949

## ON **DESIGN**

In their prolific career as furniture designers, Charles and Ray Eames were keen to explore the potential of modern materials – few of their other pieces were as traditional in style as the 670 chair, with its rich rosewood and leather. They made extensive use of cast aluminium, wooden dowels for chair and table legs, and steel rods to support chairs. The couple were also pioneers with regard to fibreglass and plastics, seizing on their potential for moulding to make a chair's seat, back, and arms in a single piece. They liked to use colourful panels in other types of furniture, integrating them into wooden display units, for example. Some of these units – made for Herman Miller (see p.109), like much of their work – combine conventional wooden drawers with perforated metal and polychrome panels. These colourful effects, together with the organic forms and narrow legs of some of their chairs, helped Charles and Ray Eames to define the look of many mid-century interiors.

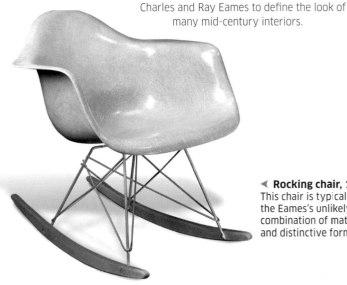

**◄ Rocking chair, 1950**
This chair is typical of the Eames's unlikely combination of materials and distinctive form.

# Wall clock

1957 ■ PRODUCT DESIGN ■ ALUMINIUM AND OTHER METALS ■ GERMANY/SWITZERLAND

SCALE

## MAX BILL

**The wall clock designed** for the German clockmaker Junghans by the Swiss designer Max Bill in 1957 is a good example of the way in which an object can encapsulate the design ideals of an entire generation. Max Bill, founding director of the Ulm School of Design in Germany, was the leading exponent of the idea of "good form". This was a postwar reinvention of the earlier Modernist notion that form follows function. In the view of Bill and his influential contemporaries, the designer had a moral and social responsibility to create functional objects that combined simplicity, quality, and durability. The wall clock exemplifies this philosophy, paring down the design to produce something that is both very simple and beautifully made.

There are no numerals on the clock face – only short lines for the minutes, and longer ones for the hours. The hands are simple metal bars, and the face is framed by a plain, light aluminium rim. In a design that is even more spartan than those produced by Le Corbusier (see p.57) and the other great Modernists of the 1930s, the clock has been reduced to the barest essentials, yet without sacrificing legibility. Bill's minimalist wall clock has influenced the appearance of countless clocks and watches in the half century since it was launched, and remains a symbol of the purity of 1950s Swiss and German design.

## Visual tour

**KEY**

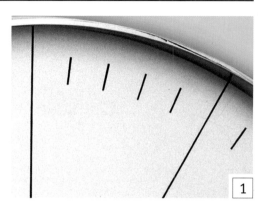

1

▲ **LINES AND RIM** The visual language of the clock is very minimal. The edge of the clock's body reads as a simple line of aluminium, while the marks indicating the hours and minutes are just bold enough to be readable against the plain white background of the clock face.

2

◄ **HANDS** The chrome-plated metal bars that make up the hour and minute hands are elegantly shaped. Their shiny surfaces give them a light appearance in keeping with the minimalist aesthetic of the clock and they are very easy to read. Their pivot point is covered with a discreet, polished-metal circle.

---

### MAX **BILL**

#### 1908-94

After an apprenticeship as a silversmith, Max Bill enrolled at the Bauhaus (see p.46) in 1927, then worked in Zürich as an architect, designer, and artist. In the 1930s, through his architectural and design work, he became known for advocating art based on rational and mathematical principles. After World War II, Bill turned to product design, and in 1950 he left Switzerland for Germany to become director of the Bauhaus-influenced Ulm School of Design. Bill also designed the school's buildings. While continuing to work as an architect, he designed furniture and lighting, and made paintings and sculptures.

### ON **DESIGN**

In the 1930s and 1940s, Swiss designers such as Max Bill developed a distinctive modern style of graphic design that relied on strong sans serif typefaces (in particular an 1890s font called Akzidenz-Grotesk), asymmetrical layouts, abstract shapes, and careful visual organization using underlying grids. In this disciplined style, designers usually restricted themselves to a single typeface in one or two weights. Through Bill and the magazine *Neue Grafik*, the style spread beyond Switzerland in the 1950s and 1960s and became known as the International Typographic Style.

► **Book cover with typography** Max Bill, 1934

"The mathematical approach in art is not actual mathematics. It is the design of rhythms and relationships"

**MAX BILL**

The long minute hand almost reaches the short marks that indicate the minutes

The hour marks are just long enough to make a continuous line with the short hour hand

JUNGHANS

The subtle aluminium rim complements the plain, simple face

# Helvetica typeface

1957 ■ GRAPHICS ■ SWITZERLAND

## MAX MIEDINGER

**In 1957, a typeface appeared** that was to become one of the most popular ever produced. Helvetica is now everywhere – on street signs, packaging, and computer screens. Its story began in the mid-1950s, when early sans serif typefaces, then generally known as grotesques, were popular with many printers and graphic designers. These letterforms, which had no serifs (terminal strokes), caught the imagination of the designers of the Modern Movement, who prized their elegant simplicity and clarity. One of the most popular was Akzidenz-Grotesk, produced by Berthold, a German type foundry. Eduard Hoffmann, President of the Swiss Haas foundry, wanted lettering that could compete with Akzidenz-Grotesk, and commissioned Max Miedinger to design a new typeface.

Miedinger's new font was unassuming but distinctive. The height of letters such as the lower-case e (known to typographers as the x-height) was generous, the strokes terminated either horizontally or vertically, and the font was set with very close spacing between the letters. This spacing made the typeface feel dense and created a strong impact, but the large x-height kept the letters legible. Miedinger and Hoffmann called the typeface Neue Haas Grotesk, and it was a success with printers and designers. Hoffmann then licensed it to the Linotype company, producers of widely used typesetting machinery, and decided to give this version of the typeface a new name. He considered Helvetia – an allegory of Switzerland – but this was already used by other companies, so he settled on the variant, Helvetica. This name clearly indicated the origin of the font and also linked it with the modern or "Swiss" typography that had such a great influence on the graphics of the time. Helvetica soon became the name for all the versions of the font. Neutral, legible, and versatile, this clean-cut typeface has now become part of daily life.

## "Helvetica is the jeans, and Univers the dinner jacket. Helvetica is here to stay"

**ADRIAN FRUTIGER**

---

### MAX **MIEDINGER**

#### 1910–80

Born in Zürich, Max Miedinger trained as a typesetter, then went to evening classes at the Kunstgewerbeschule (school of arts and crafts) there. He worked as a typographer in the advertising studio of the Globus department store in Zürich for ten years before joining the Haas foundry in Basel as a sales representative. In 1956, he went freelance as a designer of typefaces. Apart from Helvetica and its variant forms, Miedinger's best known typefaces include Pro Arte (a design that was influenced by 19th-century playbills) and Horizontal (a typeface that features ultra-wide capital letters with thin strokes).

### IN **CONTEXT**

At the same time that Miedinger was working on the font that became Helvetica, Swiss typographer Adrian Frutiger was developing another typeface. This was Univers, created for the Deberny & Peignot type foundry in Paris. Both Helvetica and Univers are based on earlier grotesque fonts but, although they share the same overall aesthetic, the letters in Univers have more classical proportions.

The typefaces also differ in many details. In the lower-case a, for example, Univers lacks the curving tail of the a in Helvetica and Akzidenz-Grotesk. For Univers, Frutiger also developed an ingenious numbering system to indicate the different weights of the typeface. This offered designers greater precision than the traditional descriptions, such as "light" and "bold".

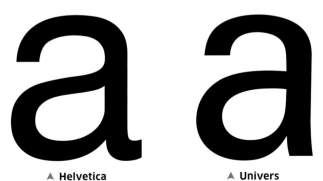

▲ Helvetica          ▲ Univers          ▲ Akzidenz-Grotesk

ABCDEFGHI
JKLMNOPQ
RSTUVWXYZ
abcdefghijkl
mnopqrstuv
wxyz123456
7890({@£$&
¥/!*°%ßfiØŒÐ?æ

# Visual tour

ABCDEFGHI
JKLMNOPQ
RSTUVWXYZ
abcdefghijkl
mnopqrstuv
wxyz123456
7890({@£$&
¥!*%ß fiØÐ?æ

**KEY**

▶ **LOWER CASE G** The bowl of the lower case g is large and occupies the whole of the letter's x-height. Helvetica's generous x-height is one of its most distinctive features, and many people find that it makes the typeface legible and easy on the eye. Another key feature of the g is the way the tail terminates horizontally; this is also the case with the c, e, and s.

1

▼ **CAPITAL R** The leg of the upper-case R curves where it joins the bowl, while maintaining a uniform stroke width from top to bottom. The leg also has a tiny curve at the base. These features contrast with the same letter in Akzidenz-Grotesk, which has an R with a straight, diagonal leg.

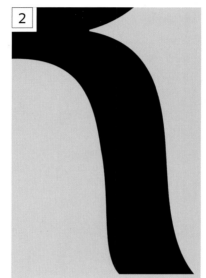

2

3

▲ **CAPITAL G** The horizontal bar of the capital G extends further to the left than it does in many fonts. It is similar in length to the downward-pointing stroke or spur, and the resulting right angle gives the letter a strong geometric character.

5

▲ **NUMERAL 1** As in many typefaces, the top of this numeral has a small stroke, so that readers can distinguish it easily from the lower-case l. In Helvetica, this stroke has a curve, a feature it shares with the 1 in Akzidenz-Grotesk. Many other sans serif typefaces, on the other hand, have a plain, straight diagonal stroke here.

4

◀ **EXCLAMATION MARK** The exclamation mark is made up of straight lines that taper gradually towards the base. The dot is a perfect square and matches the square full point and square dots on Helvetica's j and i.

# Visual tour

**KEY**

➤ **FORK HEAD** Many forks in use in Europe in the 1950s had the long, narrow tines of 19th-century cutlery. Raacke adopted shorter tines and a more generous area of solid metal beneath them. This combination makes the item easy to use, and the gently rounded shape harmonizes with the other pieces in the set.

➤ **MONO LOGO**
In lower case sans serif lettering, the Mono logo was an important part of the design. The name appeared in large type on the packaging and was also positioned fairly prominently on the knife blade. Users instantly recognized the brand, and this increased its success.

➤ **HANDLE** The Mono series has straight handles. Although they look austere compared with traditional cutlery, these handles are meticulously finished with slightly curved edges and faultless surfaces. The numerous grinding and polishing processes involved in producing these perfectly shaped pieces makes them comfortable to hold, and their unadorned appearance gives the range its timeless quality.

mono INOX SOLINGEN GERMANY

The immaculately finished spoon required more than 20 production steps

The uniform thickness of the handle results from the use of a standard gauge of sheet metal

**Spoon**          **Teaspoon**

## ON **DESIGN**

Raacke's original idea was to produce a set of flatware that was so perfect on its own terms that no other ranges would be necessary. Subsequently, however, he produced several variations. A children's set appeared in 1959 and Raacke also designed versions with ebony and teak handles, which some people felt offered a better grip. These ranges were designated Mono-e and Mono-t, and the original line became known as Mono-a. A few years later, in 1962, the company added the Mono-Ring, a design that had a plastic handle and terminal ring so the pieces could be hung up. Mono-Oval, which had heavier, oval-profile handles in stainless steel, was launched in 1982.

➤ **Mono-Ring flatware**, Peter Raacke, 1962

# Austin Seven Mini

1959 ▪ CAR DESIGN ▪ STEEL ▪ UK

SCALE

## ALEC ISSIGONIS

**The Mini was a radical design** that transformed motoring, and the box-shaped car became the symbol of an era. Its development was the result of the search for a small, economical car during the fuel shortages after the Suez Crisis of 1956. At this time, bubble cars imported from Italy and Germany, such as the VW Beetle (see pp.80–83), were becoming popular, and British car manufacturer BMC (created by a merger of the Austin and Morris companies) decided to offer UK customers a home-grown alternative. The company's chief engineer, Alec Issigonis, rethought the small car, doing everything he could to cram four people, their luggage, and a four-cylinder engine into the smallest possible package – a body just 3 metres (10 feet) long. A short front end and minimal boot cut down the length, and

by using front-wheel drive and smaller wheels (and wheel arches), Issigonis saved space in the interior. Every bit of spare room was exploited, from a storage area under the rear seat to storage bins in the doors. Decorative trim was kept to a minimum to emphasize the functional design.

Although sales were initially slow, the car took off in the 1960s. It combined charm, good performance (proved by its rallying successes), fuel economy, and pared-down design, and found favour with both fashion-conscious and cost-conscious buyers. The Mini became an enduring emblem of the 1960s, and production continued until the year 2000, by which time more than five million cars had been made worldwide. In 2001, BMW launched the new-generation Mini Hatch.

### ALEC **ISSIGONIS**

#### 1906–88

Born in Turkey, Alec Issigonis came to the UK as a teenager and attended Battersea Polytechnic, where he studied engineering. He was a skilled draughtsman and worked in the drawing office at Humber cars before moving to Morris in 1936, where he helped develop independent suspension and rack-and-pinion steering systems. He was involved with the design of many cars for Morris and later BMC, including the hugely popular Morris Minor in the late 1940s, and the 1100 and Maxi in the 1960s. Issigonis became Engineering Director at BMC in 1964 and received a knighthood in 1969.

▲ **Alec Issigonis** standing alongside a Morris Mini at the Fighting Vehicles Research and Development Establishment, 1959

The bonnet opens from the front

The curve of the front windscreen echoes that of the rear window

The sills were later redesigned to minimize water damage

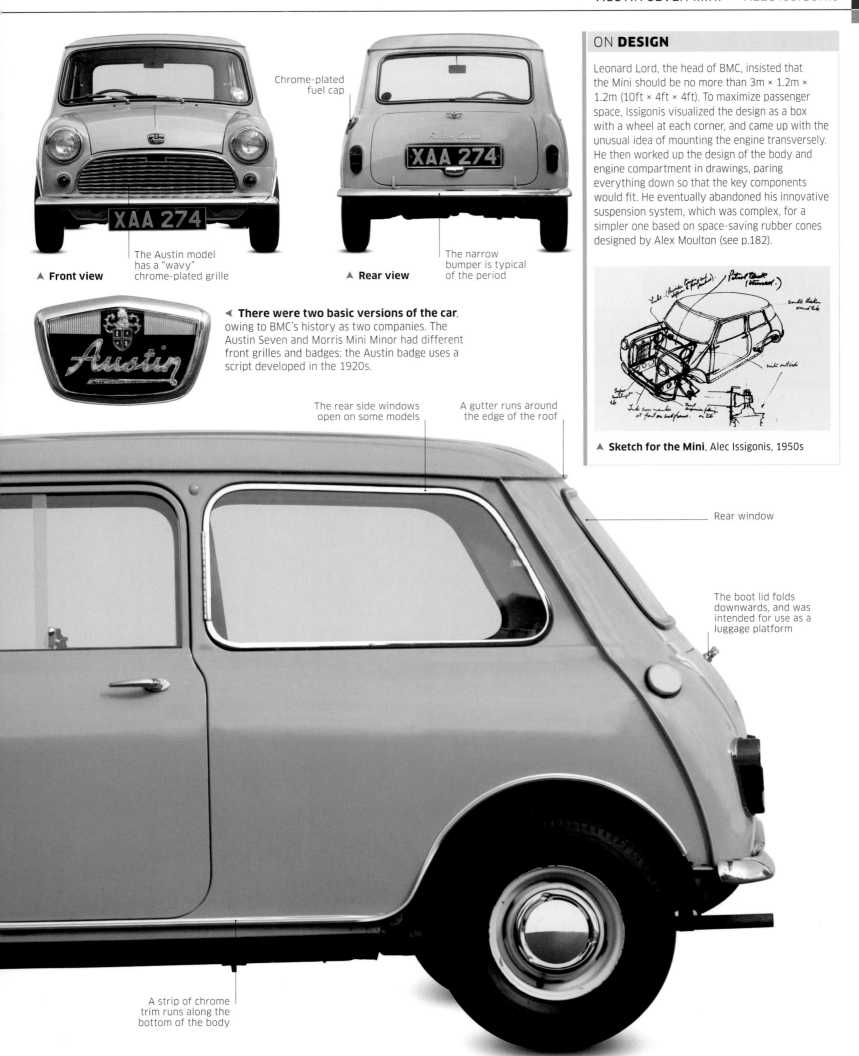

**▲ Front view**

Chrome-plated
fuel cap

The Austin model
has a "wavy"
chrome-plated grille

**▲ Rear view**

The narrow
bumper is typical
of the period

**◄ There were two basic versions of the car**, owing to BMC's history as two companies. The Austin Seven and Morris Mini Minor had different front grilles and badges: the Austin badge uses a script developed in the 1920s.

## ON DESIGN

Leonard Lord, the head of BMC, insisted that the Mini should be no more than 3m × 1.2m × 1.2m (10ft × 4ft × 4ft). To maximize passenger space, Issigonis visualized the design as a box with a wheel at each corner, and came up with the unusual idea of mounting the engine transversely. He then worked up the design of the body and engine compartment in drawings, paring everything down so that the key components would fit. He eventually abandoned his innovative suspension system, which was complex, for a simpler one based on space-saving rubber cones designed by Alex Moulton (see p.182).

**▲ Sketch for the Mini**, Alec Issigonis, 1950s

The rear side windows
open on some models

A gutter runs around
the edge of the roof

Rear window

The boot lid folds
downwards, and was
intended for use as a
luggage platform

A strip of chrome
trim runs along the
bottom of the body

# Visual tour

**KEY**

▼ **FRONT LIGHTS** The arrangement of the front lights could not be simpler, with the single headlamps and conical direction indicators mounted separately, one above the other. This same basic arrangement was used on all the early models, but there were variations in the grille design.

➤ **WHEELS** The Mini has wheels with a diameter of just 25.4cm (10in) although Issigonis suggested at one point that they should be 20.3cm (8in). Having small wheels means that there is a modest amount of additional passenger space inside, It increases the available luggage storage at the back of the car and makes the engine compartment larger at the front.

▼ **SLIDING SIDE WINDOW** Unlike most car windows, the Mini's front windows slid from side to side when released by a simple catch inside. There was therefore no need for a window-winding mechanism and the door could be made of a single sheet of metal, instead of two.

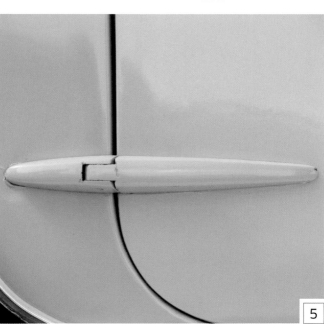

▲ **DOOR HANDLES** The door handle is a very simple, forward-facing design, tapered to make it look aerodynamic. On later models, the handles were flattened and partially recessed into the door panel, not only to give the outline a sleeker appearance, but also to avoid injury to pedestrians.

◄ **DOOR HINGES** Among the most stylish elements of the Mini design are the streamlined, torpedo-like door hinges, which are finished in the same colour as the rest of the bodywork. Intended by Issigonis to be purely functional, these hinges also help to make the car different in appearance from the other small cars produced at the time.

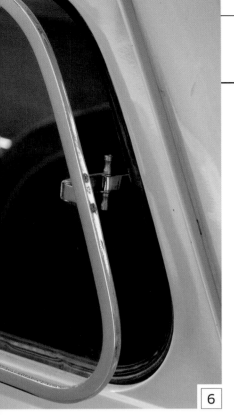

◄ **HINGED REAR WINDOW** Hinged on the central pillar, the rear side windows open outwards to allow a small amount of additional ventilation into the rear part of the interior. The fastener, which holds them in position, avoids the need for a winder mechanism, and saves valuable space.

▼ **BOOT LID AND NUMBER PLATE** The downward-opening boot lid doubled as a luggage rack, and on early models the number-plate panel was also hinged so that it could be visible when the boot lid was open. This feature was later removed when the manufacturers realized that toxic exhaust fumes could travel through the open boot into the interior of the car.

6

7

► **SPEEDOMETER** The circular speedometer was the only instrument on early models and was set in the middle of the dashboard, to one side of the steering wheel. This was outside the driver's normal line of vision, but the large dial and numerals made it easy to read at a glance. Also incorporated into the dial were the odometer, fuel gauge, and warning lights for ignition, oil pressure, and the headlamp beam.

8

## ON **DESIGN**

The basic Mini had an 848cc engine, which was small but powerful enough for such a lightweight car. Mounting the engine transversely (with its long axis parallel to the car's axles) was an ingenious space-saving measure. Issigonis did not invent the transverse engine – it had been used before by several manufacturers, including DKW and Borgward in Germany and Saab in Sweden. As well as reducing the length of the front of the car, a transverse engine can drive the front wheels, saving room occupied by the drive shaft – a necessary feature on rear-wheel drive cars. Tucked away under the engine was the gearbox, again taking up less space than in a conventional car.

▲ **Transversely mounted engine**

## IN **CONTEXT**

The Mini soon spawned a number of variants with different engines and body shapes. One of the most successful was the sporty Mini Cooper, a high-performance version originally with a larger 997cc engine, front disc brakes, and other improvements. It won many rally victories. Commercial vehicle designs included a van that proved popular with a wide range of users from small businesses to police forces, and a modest-sized pick-up. Domestic users who required more flexible carrying capacity tended to opt for one of the estate car versions with rear double doors . The version with non-structural wooden trim was especially popular and variously badged as the Austin Mini Countryman or the Morris Mini Traveller. A further model was the Moke, an open-topped utility vehicle like a miniature Jeep. Originally intended for military use, it proved popular in a variety of leisure contexts.

▲ **Mini Cooper**

▲ **Mini Moke**

▲ **Mini Traveller**

▲ **Mini Pick-up**

# Cadillac series 62

1959 ■ CAR DESIGN ■ VARIOUS MATERIALS ■ USA

SCALE

## HARLEY EARL

**During the first half of the 20th century**, Cadillac established itself as a successful upmarket marque in the US car industry. Cadillacs were known for their luxury, comfort, size, range of features, and, thanks to designer Harley Earl, their distinctive styling. In 1959, the company launched a new batch of models in their long-running 62 series, and these vehicles, with their elongated, low-slung bodies and enormous rear tail fins, became the most recognizable American cars of the period.

The relaunched series 62 range included hardtops and the sleek, two-door convertible, which was a staggering 6.1 metres (20 feet) long. All the cars featured power-assisted brakes, power steering, automatic transmission, and two-speed wipers, while the big 6.4-litre engines provided the smooth performance that had become the norm for Cadillacs. The stylish interiors had big, adjustable leather seats and colour schemes that matched the exterior paintwork. What set the 1959 cars apart, however, was their flamboyant body styling.

The Cadillac's most distinctive features, the rear tail fins, were an adornment that Earl had introduced in 1948. Other US manufacturers, such as Chrysler, followed suit, and there was a burgeoning fashion for fins in the 1950s. However, none were as big as the stupendous ones on the 1959 Cadillacs, and, although some critics attacked their design as excessive, pointless, or even ugly, they caught the imagination of consumers. The fins, combined with the generous amounts of gleaming metal at both front and rear, the long, low bodies, distinctive rear lights, and wraparound windscreens, made the cars unmistakeable.

In the USA, where the country's great size, successful car industry, and inexpensive petrol had made cars a central part of the national way of life, the large, luxurious Cadillacs stood out as aspirational vehicles. Their glamour helped to keep large cars on America's roads for a generation.

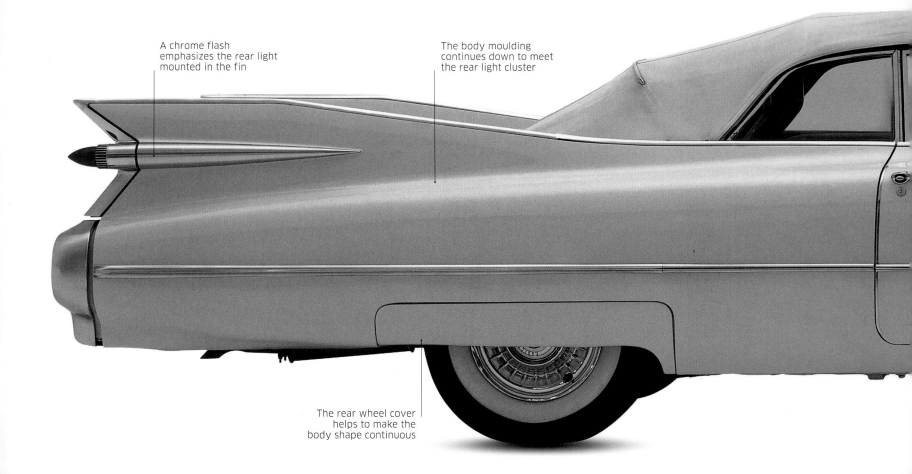

A chrome flash emphasizes the rear light mounted in the fin

The body moulding continues down to meet the rear light cluster

The rear wheel cover helps to make the body shape continuous

Twin
headlamps

The wing mirror
was a standard
feature

Twin
parking
lights

▲ **Front view**

Tail and
direction
indicator
lights

The rear trim
matches the
front grille

▲ **Rear view**

## HARLEY EARL

### 1893–1969

Harley Earl left university early to
work for his father, whose company
produced customized car bodies. He
developed the technique of using
modelling clay to sculpt life-size
bodywork designs and was spotted by
Cadillac's Laurence P. G. Fisher, who
hired Earl to design the new La Salle.
The success of this project led to Earl's
appointment as director of General Motors' Art and Colour
Section, making styling a priority in the car industry for the
first time. Later, Earl was promoted to Vice President of
the company – an unprecedented promotion for a styling
specialist. A succession of bestselling designs followed,
including the 1939 Buick Y-Job (the first-ever concept car),
the 1948 Cadillacs (the first to have tail fins), and the 1953
Chevrolet Corvette. Wraparound windows and two-tone body
paint were among his most influential innovations. He retired in
1958 when the designs for the 1959 Cadillacs were complete.

▲ **Cadillac emblem** The Cadillac emblem used on the
1959 cars combines two elements. The coat of arms belongs
to Antoine de la Mothe Cadillac, who founded Detroit in 1701
and gave the company its name. The other element, the large "V",
was used by the company to symbolize their powerful V8 engines.

Whitewall tyres were
an optional extra on
most models

The wraparound
front bumper
extends to the
front wheel arch

# Visual tour

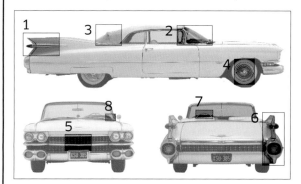

**KEY**

▶ **FIN** The tail fins of the Cadillac rise more than 1 metre (3¼ feet) above the ground, doubling the height of the rear of the vehicle. Although some believed that tail fins improved a car's aerodynamic performance, their main purpose was to look extraordinary. The way in which they come to a point suggests arrow-like speed, while their general shape recalls the tail fins of aircraft, one of Earl's sources of inspiration.

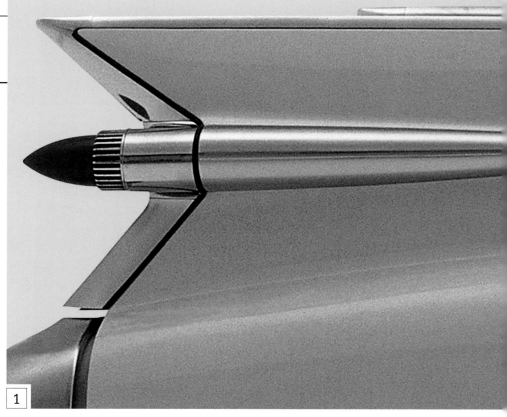

1

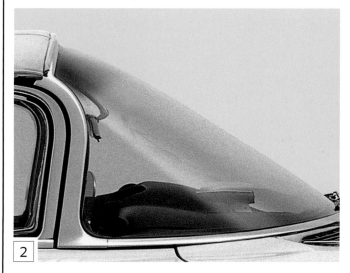

2

5

▲ **WINDSCREEN** The Cadillac's wraparound windscreen was a common feature of General Motors cars of the period, including Buicks and Chevrolets. The way in which the windscreen curved around the side of the vehicle provided good visibility for the driver. The curved steel frame is another stylish detail.

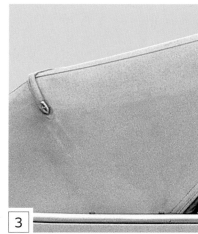

3

▲ **SOFT TOP** The soft top on this car is in a cream-coloured fabric that harmonizes well with the pink paint finish. Most soft tops of the period had dark fabric, so this was another styling detail that set the car apart.

4

▲ **RADIATOR GRILLE** The eye-catching front grille is covered in chromed metal and was designed to draw attention to the car as it approached. The upper section makes a shining grid pattern and is divided from the lower part of the grille by a horizontal bar. This bar comes to a point at the centre, so that it forms a shallow V-shape, reinforcing the "V" used in the Cadillac emblem.

◀ **WHEEL AND TYRE** The large hubcaps are embossed with a pattern that resembles radiating spokes. The tyres on this model have white walls, which were a popular option on cars during the 1950s and 1960s. Car buyers liked the way they blended with the pale body colours that were becoming increasingly popular.

7 ◀ **STEERING WHEEL AND DASHBOARD** The interior layout is high-spec and luxurious. Cadillac paid particular attention to the colour scheme, using the same pink as the bodywork on the steering wheel. Pale leather upholstery harmonizes with the colour of the soft top.

▼ **POWER WINDOW CONTROLS** This model has electric windows, controlled from the driver's door. The bright metal switches are set in a cream panel, to blend with the colour of the seats and the lower section of the dashboard.

8

◀ **REAR LIGHT CLUSTER** The lower rear light setting was designed to integrate with the bright metal bumper. It features a central, white reversing light surrounded by red brake lights. Above, halfway up the tail fin, are twin rear lights in a pair of little pods that look like a science-fiction illustrator's idea of space rockets or jet engines.

6

## ON **DESIGN**

Cadillac produced the series 62 in a range of body styles that included a coupe that had two doors and a downward-sloping roof, contrasting with the straight roof of the base-model sedan. One of the most sought-after models was the Coupe de Ville, which combined an impressive equipment specification with the elegant coupe body. All series 62 models display the huge tail fins with their distinctive rear lights, and the low-slung bodies are similar – the look was said to be an attempt to outclass Chrysler, whose 1950s models had been received particularly well. Although attributed to Harley Earl, a team of designers was involved, and several of the details, including the tail lights, may have been the work of other members of the group.

## IN **CONTEXT**

One of the inspirations for car tail fins was said to be the rear of the Lockheed P-38 Lightning fighter aircraft, first manufactured in 1941. Harley Earl added similar fins to the 1948 Cadillacs; they were later taken up by Virgil Exner, who designed successful 1950s Chryslers with larger fins, while the 1959 Cadillacs had still bigger ones. The fact that these cars bore some resemblance to an aircraft was hugely important to their sales – people bought costly vehicles like the Cadillac for their luxury and because the cars had an image that combined power and adventure. The aeronautical link brought to the car some of the allure of the world of the fighter pilot, and turned the Cadillac into a dream car for many Americans.

▲ **Cadillac Coupe de Ville**, 1959

▲ **Lockheed P-38 Lightning aircraft**, 1941

# Panton chair

1959-60 ■ FURNITURE ■ PLASTIC ■ DENMARK

SCALE

## VERNER PANTON

**The first chair to be made from a single piece** of plastic, the Panton chair marked a turning point in 20th-century design. It was the brainchild of designer Verner Panton, and its futuristic form caused a sensation. Panton began to experiment with single-form chairs in the 1950s, when he created the S chair, made of bent, laminated wood and inspired by Gerrit Rietveld (see p.34). By 1960, after a visit to a factory that made plastic buckets and helmets, Panton had come up with the idea of making a similar chair in plastic. His cantilevered chair closely followed the contours of the seated human body, and its serpentine shape and backward-sweeping base created a strong and precisely balanced structure. It was an ingenious design, but there was a problem – the chair was difficult to make, and finding the right material was not easy. The German

manufacturer, Vitra, took up the challenge in 1967, making prototypes and a small production series in fibreglass-reinforced polyester. The following year, the first successful mass-production runs of the Panton chair used a rigid polyurethane foam, injected into a mould. Later runs utilized other types of plastic, latterly polypropylene, which has the right balance of strength and flexibility.

With its distinctive, sculptural shape and bright colour, the Panton chair became an emblem of the Pop era and was very successful in the late 1960s and early 1970s. More recently, following a famous 1995 *Vogue* magazine cover featuring the model Kate Moss on a Panton chair, interest in the designer's work was rekindled. The chair went back into production and is now considered to be a milestone in modern design.

### VERNER **PANTON**

#### 1926-98

Danish-born designer Verner Panton was educated in Odense and at the Royal College of Art in Copenhagen, where he was taught by Poul Henningsen (see p.164). In the early 1950s, he was an assistant to Arne Jacobsen (see p.162), before going freelance and creating pieces such as the Cone chair (1959). In the following years, he designed the first inflatable furniture and brightly coloured Op art interiors that he dubbed "total environments". In the 1960s, his use of bright colours and new materials placed him at the forefront of Pop design. His collaboration with Vitra led to the Panton chair and other furniture, as well as lighting for Poulson.

## Visual tour

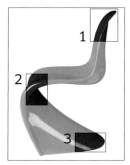

**KEY**

▶ **BACK** The back of the chair terminates in a slight backward tilt. The edges curve slightly inwards at the sides, so that the top of the chair seems to taper away to almost nothing when seen in profile. The sense of lightness that this conveys forms a contrast to the comparative solidity of the chair's base.

1

▶ **CURVING SIDES** The plastic wraps inwards at the edges of the chair, to form sides that give the chair additional strength and rigidity. This effect is particularly pronounced where the base merges with the seat. The chair needs this degree of reinforcement to support the weight of the person sitting on it.

2

3

▲ **BASE** The ingeniously designed base sweeps back so that it extends almost to the full depth of the chair. This generous base gives the chair stability and makes it visually pleasing, as the curved rear edge matches the curves above.

## IN **CONTEXT**

In the 1960s, several designers, including Verner Panton and Eero Aarnio, were using synthetic materials, such as plastics, and new production techniques to create furniture such as the Ball chair (right) in unusual shapes and colours. These designers mostly saw their work as belonging to the Modernist tradition, in which form follows function. The Panton chair, the Ball chair, and Panton's stunning Cone chair, however, also fitted easily into the aesthetics of Pop art – a movement that was a reaction against the purity and restraint of Modernism. In the 1960s, designers created objects that were playful, brightly coloured, and inspired by popular culture. Scientific advances also exerted a strong influence: Panton's furniture was considered futuristic in form and Aarnio's Ball chair resembled a space capsule.

▶ **Ball chair**, Eero Aarnio, 1963

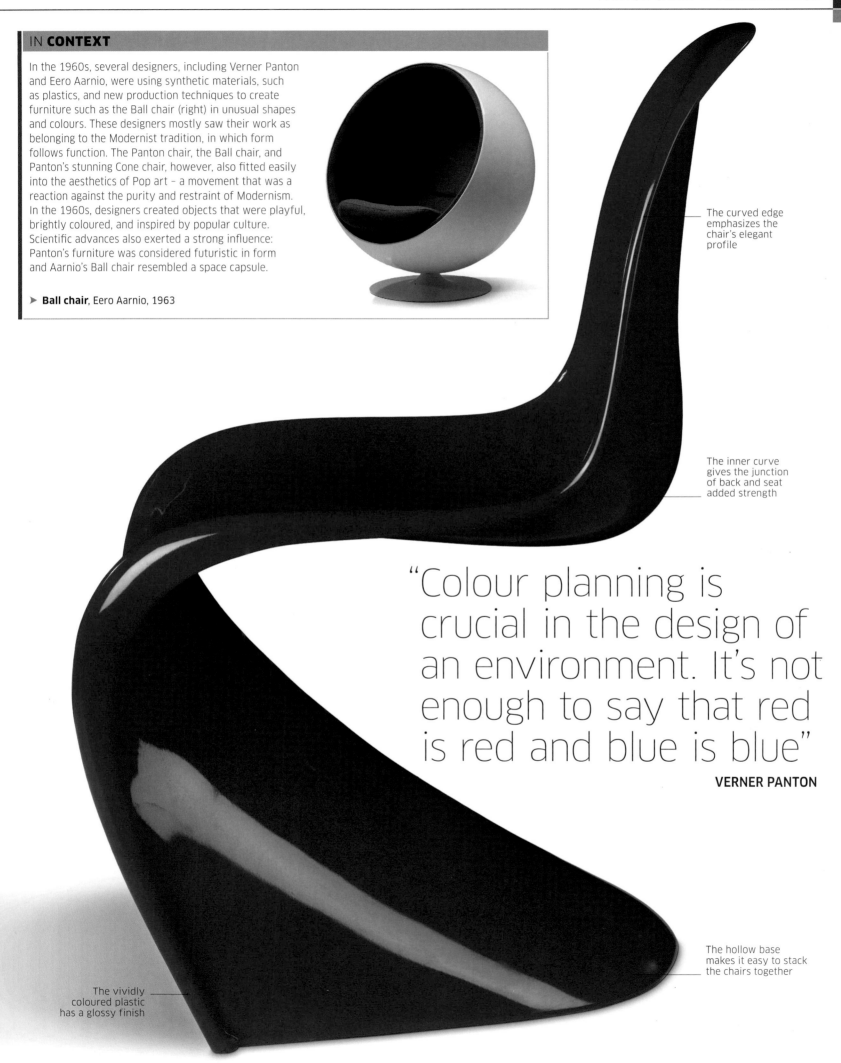

The curved edge emphasizes the chair's elegant profile

The inner curve gives the junction of back and seat added strength

"Colour planning is crucial in the design of an environment. It's not enough to say that red is red and blue is blue"

**VERNER PANTON**

The hollow base makes it easy to stack the chairs together

The vividly coloured plastic has a glossy finish

- **Arco lamp** Achille and Pier Giacomo Castiglione

- **Moulton bicycle** Alex Moulton

- **Unikko fabric** Maija Isola

- **Grillo folding telephone** Marco Zanuso, Richard Sapper

- **Kodak Instamatic 33** Kenneth Grange

- **Valentine typewriter** Ettore Sottsass

- **Revolving cabinet** Shiro Kuramata

- **Beogram 4000** Jacob Jensen

- **Tizio desk lamp** Richard Sapper

- **Wiggle chair** Frank Gehry

- **Munich Olympic Games pictograms** Otl Aicher

- **Suomi tableware** Timo Sarpaneva

1960–1979

# Arco lamp

1962 ▪ LIGHTING ▪ STAINLESS STEEL AND MARBLE ▪ ITALY

SCALE

## ACHILLE AND PIER GIACOMO CASTIGLIONI

**The great Italian designer Achille Castiglioni** believed that his role was to find elegant and functional solutions to the difficulties and challenges of creating products. One design problem that he addressed with his brother, Pier Giacomo, was how to light a dining table without drilling holes in the ceiling or leaving wires trailing. His solution was Arco (Italian for "arc" or "arch"), one of the simplest and most beautiful pieces of lighting design of the 20th century.

It is said that the Castiglioni brothers were inspired by the structure of street lamps, which use a tall column and a straight or arch-like arm to spread their light over a wide area. For the home, the brothers translated this utilitarian piece of street furniture into a perfectly balanced, high-quality floor lamp. Its slender steel upright supports a dramatic arch that curves some 2.5 metres (8ft) above the floor, with the lampholder and shade held about 2 metres (6ft 6in) from the base. The height and span of the support make it possible to place a large table under the light, while leaving enough room for the diners. The steel arch, which also conceals the lamp's wiring, forms a stunning, semi-circular sweep, and is also adjustable. The curve is actually made up of three sections that telescope into one another, so that the position of the lamp can be adjusted according to the size of the room or table.

Another key element in the design is the lamp's base. This had to be heavy, to counterbalance the arch and shade, and the Castiglioni brothers hit upon the idea of forming it out of a solid block of marble, adding a touch of luxury to their design. The combination of shade and support in polished steel with a base in Carrara marble was unusual – steel is a carefully engineered industrial material, and marble a natural substance traditionally carved into shape. However, both can have immaculate polished surfaces and the use of marble, the material most prized by Renaissance sculptors and architects, turned a functional piece of design into something that was not only stylish, but decidedly glamorous. Not surprisingly, the lamp was soon in demand by upmarket interior designers, and Flos, the lighting company that manufactured the piece, had a success on their hands. The Arco floor lamp sold widely and was much copied for its elegance and simplicity. As a design, it has stood the test of time, and remains one of the most familiar examples of Italian design flair of the 1960s.

---

### ACHILLE AND PIER GIACOMO **CASTIGLIONI**

#### 1918-2002 AND 1913-68

Italian brothers Achille and Pier Giacomo Castiglioni both studied architecture in Milan. They began working together in 1945, although Pier Giacomo had previously teamed up with a third brother, Livio, and the three continued to collaborate occasionally and to work with other designers. Achille and Pier Giacomo were among Italy's most prolific and influential industrial designers, creating a number of different lamps for Flos and other clients, a vacuum cleaner for Rem, and seating for Zanotta. Commissions for their designs came from such internationally renowned companies as Kartell, Knoll, Lancia, and Siemens. In addition to their work in the studio, they were prominent in the design community in Italy, founding the *Associazione per il Disegno Industriale* (Association for Industrial Design). The two brothers also taught architecture, and Achille became head of the architecture faculty at the Politecnico in Milan, where he had been a student. He continued to work independently after Pier Giacomo's death, and his innovative designs ranged from oil and vinegar flasks with hinged lids for Alessi to cutlery, glassware, and furniture.

▶ **Achille** (left) **and Pier Giacomo** in their studio

The curved support describes an arc or semicircle

## Visual tour

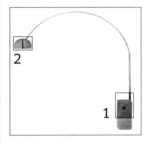

### KEY

➤ **BASE** The marble base is finely finished with chamfered edges, as if a sculptor has just started to work on it. One end of the fixing that holds the upright firm is inside it. The most striking feature of the base is the large central hole on the side. This is just big enough for a broom handle to be pushed all the way through it, making it possible to move the heavy lamp.

**1**

**2**

▲ **SHADE** The sculptural, polished steel shade, which has two parts, provides a focal point above a table. The cap, which is fixed to the lampholder, has a group of holes carefully arranged so that the heat from the standard incandescent bulb can escape. The outer ring of the shade (the reflector) is moveable, so that its position and angle can be adjusted easily.

### ON **DESIGN**

Achille Castiglioni was famous for drawing inspiration from everyday objects - for making stools from bicycle or tractor seats, and for basing the design of the Arco lamp on a street light. He knew, however, that radical ideas alone were not enough. He worked out each design painstakingly, and his meticulous drawings (see below) show how he refined the shape and size of the steel sections that made up the support of the Arco lamp. Each piece had to be just the right size to fit inside the adjoining section, with the narrowest piece at the end nearest to the shade. Castiglioni also designed a series of stops, so that each piece clicked into place securely, leaving enough room for the wire inside.

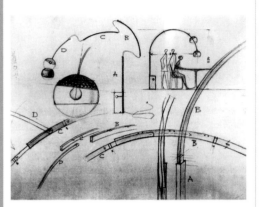

▲ **Working drawing** of the components of the Arco lamp, 1962

The upright is made of immaculately finished, square-section stainless steel

The marble base, weighing about 65kg (143lb), provides stability

"Start from scratch, stick to common sense, know your goals and means"

**ACHILLE CASTIGLIONI**

# Moulton bicycle

1962 ▪ PRODUCT DESIGN ▪ VARIOUS MATERIALS ▪ UK

SCALE

## ALEX MOULTON

**By the late 1950s, bicycle design** had changed very little since the introduction of cycles with two large wheels in the late 19th century. The basic combination of wheels around 66cm (26in) across and a diamond-shaped frame seemed to offer very little scope for radical change. In 1956, however, when the Suez crisis posed a threat to European oil supplies, many people turned away from their cars and it seemed the right time to rethink the bicycle. Engineer Alex Moulton set himself the task and his innovations changed the face of cycling.

The bicycle Moulton developed had two distinctive features: small wheels and a step-through, F-shaped frame with no crossbar. He realized that smaller wheels have lower aerodynamic drag than conventional, large cycle wheels, and he combined them with high-pressure tyres for better performance. The bicycle was easy to steer, and the small wheels gave it a lower centre of gravity, which made it safer to ride. The combination of small wheels and high-pressure tyres, however, could also lead to a rough ride, so Moulton drew on his experience of car design and built suspension into his prototype from the start. The result was a comfortable bicycle that was easy to ride and suitable for both men and women.

By 1962, the design was ready to launch commercially and although the bicycle's unusual appearance surprised some, others were won over by its modern look. The bike was an immediate success, but demand easily outstripped

what Moulton's small company could supply, so he set up a new manufacturing base under the auspices of the British Motor Corporation, makers of the Mini. When rival companies began to introduce small-wheel models, it was clear that the Moulton had set a new standard for cycle design.

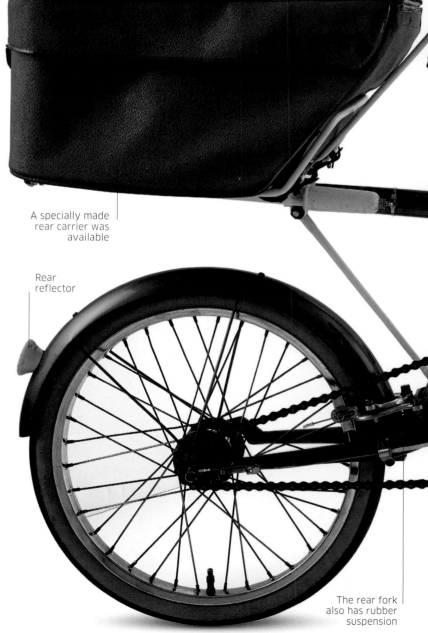

A specially made rear carrier was available

Rear reflector

The rear fork also has rubber suspension

◀ **The cycle's badge** bears the Moulton logo. Its double Ms were picked out in blue on all early F frames. The badge also features Bradford-on-Avon, Alex Moulton's home town, where the bicycle was developed and originally manufactured.

Moulton Bicycles
Bradford on Avon
England

## IN **CONTEXT**

Moulton's initial idea was to build the cycle's frame as a monocoque construction – a body made from pieces of sheet metal, riveted together. Moulton had several prototypes built in this way in the late 1950s. Some people found them similar in appearance and structure to Vespa scooters (see pp.98–101), and reminiscent of aircraft design, with which Moulton was very familiar. These prototypes, which were made of lightweight aluminium, looked attractive and seemed promising, but the monocoque shell made a loud noise when ridden, so Moulton abandoned it in favour of rigid steel tubes.

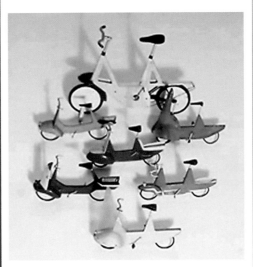

▲ **Monocoque prototypes,** Alex Moulton, 1958

The height of the handlebar is adjustable

A front light can be mounted on this bracket

## "The greatest work of 20th-century British design"

**NORMAN FOSTER**

The frame is made of large steel tubing, for strength and rigidity

This horizontal section acts as a carrying handle

Known as the F-frame, the long main section has two branches that support the handlebars and saddle

Moulton deluxe

High-pressure tyre

# Visual tour

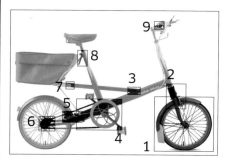

**KEY**

▶ **WHEELS** The Moulton's wheels are just 40.6cm (16in) in diameter, less than two-thirds the size of conventional bicycle wheels. When fitted with high pressure tyres, they roll even more efficiently than standard types. Small wheels accelerate faster than larger ones and are stronger and easier to steer.

▼ **SUSPENSION** The Moulton's distinctive front suspension, designed to compensate for the rigid frame and small wheels, smoothes out the ride on bumpy road surfaces. Drawing on his experience designing car suspension systems, Moulton used an internal rubber spring, and a metal coil-spring covered by an outer rubber bellows.

1

2

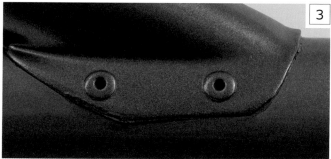

3

◀ **FRAME CONSTRUCTION**
Conventional bicycle frames were made by joining lengths of tube together with metal sleeves called lugs, a technique particularly appropriate for tubing that was narrow in diameter. Moulton created his frame with wider tubing held together with pop rivets and then brazed (a technique similar to soldering). This method, which was used in the aircraft industry, created a more rigid frame.

4

▶ **CHAIN GUARD** The white plastic guard on the chain wheel is one of the most distinctive features of the Moulton. It prevents the rider's clothing from coming into contact with the chain and gives the bicycle a clean, modern look.

5

## ON **DESIGN**

The success of the Moulton was partly due to its practicality and comfortable ride, but many owners also appreciated Moulton's huge achievement in radically rethinking bicycle design. The Moulton seemed to epitomize the modern look, and became the bicycle of the "Swinging Sixties", appearing in numerous films and fashion shoots. The Standard, Deluxe, Speed, Stowaway, and Safari models all appeared at the launch at Earl's Court, London in 1962. Three years later, Moulton launched the Speedsix, a racing and touring small-wheel bicycle that was the first mass-produced cycle of any type with a six-speed derailleur gear.

▲ **Bicycle frame joint**

▲ **Moulton Stowaway**, Alex Moulton, 1965

▲ **REAR BRAKE**
This Deluxe model has efficient caliper brakes at the front and rear, with cables connecting the calipers to the brake levers on the handlebar. On the Stowaway model, which could be taken apart for storage, the rear caliper was replaced with a hub "coaster" brake, which meant that there was no need for a rear brake cable.

◄ **REAR HUB** The original Moulton Deluxe had internal hub gears (a system in which the moving parts are within the wheel hub) offering four speeds. The rider changed gears using a four-speed thumb-controlled trigger on the handlebar.

6

8

◄ **SADDLE ADJUSTMENT** It is quick and easy to adjust the height of the saddle. Turning this lever loosens the column into which the saddle is fitted, so that the saddle can be lowered or raised to suit different cyclists. No spanner is required, which is not the case with many bicycles.

▼ **BELL** The bell is marked with the distinctive Moulton double M logo. Pressing the lever with your thumb releases the sprung clapper to create a loud, clear ping.

▶ **PUMP** To keep the Moulton's tyres inflated to the correct pressure, a bicycle pump is stowed on the frame beneath the rear carrier, clipped into place by an integral peg at the back of the carrier and a small chrome peg clamped to the frame (right). A pump was fitted to all new bicycles as standard.

7

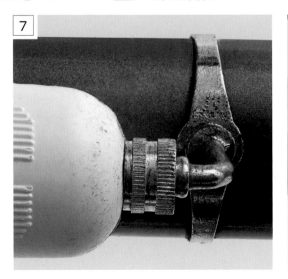

9

# Unikko fabric

1964 ■ TEXTILE DESIGN ■ SCREEN-PRINTED COTTON ■ FINLAND

## MAIJA ISOLA

SCALE

**Few designs sum up an era** as effectively as the Unikko (Poppy) print created by Finnish designer Maija Isola for the Marimekko company in 1964. Its simple, stylized design is typical of the 1960s, but the astonishing boldness of the pattern, with its big, brilliantly coloured flowers, took people by surprise.

The design very nearly did not appear at all. Armi Ratia, the visionary co-owner and creative leader of Helsinki-based Marimekko, liked strong colours, but decided that the company should not produce floral patterns – she thought that the natural beauty of flowers would be compromised by turning them into motifs on fabric. However, the strong-willed Maija Isola decided to trust her own intuition and create a floral pattern in any case. Ratia liked the design and immediately put it on sale, but the public was slow to respond. Realizing Marimekko's customers were unable to see the potential of the print, Ratia had some shift dresses made of the material and took them to a fashion show. The response was immediate: women snapped up the dresses and the print became a bestseller. The fabric was produced in different colourways, and used on bags, umbrellas, and bed linen, as well as for clothes. Marimekko became widely recognized as a major supplier of the brightest and boldest modern fabrics, and Isola's signature print remains one of the best-loved of all the company's designs.

## Visual tour

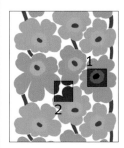

**KEY**

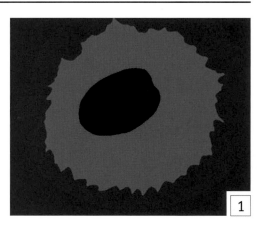

1

▲ **FLOWER CENTRE** The design is composed of confidently printed, flat flower forms without fine detail. A loose oval represents the poppy's dark centre, and the jagged orange outline suggests a frilly mass of stamens.

2

◄ **PETAL EDGES** The rounded outer petals of the flowers are painted in broad curves, and arranged quite closely but without overlapping. The dazzling combination of vibrant red and shocking pink stands out graphically against the plain white background.

## MAIJA **ISOLA**

### 1927–2001

After attending Helsinki's Central School of Industrial Arts, Maija Isola joined oilcloth company Printex, in 1949. Using silkscreen techniques, she printed her bold designs on sheets of fabric. Printex became Marimekko in 1951, and Isola continued to work for the company until 1987. By the end of her working life she had built up a body of some 500 different patterns. In the 1980s, Isola collaborated with her daughter, Kristina, a textile artist. Kristina carried on working for Marimekko after her mother's death and revived some of Maija's best-known designs by introducing new colourways (see right).

## ON **DESIGN**

Starting with designs drawn from nature (some reproduced photographically for a realistic effect) and moving on to abstract patterns, Maija Isola's work became bolder as she grew in confidence and experience. She also branched out into designing textiles with motifs borrowed from historical and traditional cultures. For some of these, Isola drew inspiration from the Byzantine art she saw when travelling in southern Europe. Others were derived from the folk art of Karelia, a forested region on the borders between Russia and Finland.

► **Alternative colourway** of the Unikko print

# Grillo folding telephone

1965 ▪ PRODUCT DESIGN ▪ ABS PLASTIC ▪ ITALY

SCALE

## MARCO ZANUSO, RICHARD SAPPER

**The Grillo telephone**, created by Italian designer Marco Zanuso and his German partner Richard Sapper, was a truly groundbreaking piece of industrial design. In the 1960s, most people were using telephones based on the Ericsson model of 1931 (see pp.68–69), with its large, dark Bakelite body and separate handset. Zanuso and Sapper, by contrast, fitted the whole appliance into one unit and, thanks to the latest electronics technology, made it less than half the size of previous telephones. The two men devised an ingenious clamshell design that opened when you lifted it to reveal the dial, earpiece, and mouth-piece. To keep the unit as compact as possible, the ringer was placed in the plug, and in another radical

move, this ringer emitted an electronic chirping noise that sounded like a cricket (*grillo* in Italian). The Grillo's other distinctive element was its body, which was made of plastic and available in a range of colours including white, red, orange, and blue. Plastics were making increasing inroads into the home in the 1960s (Zanuso and Sapper also designed a child's plastic chair) and the bright colours and curvaceous moulded shapes that the material made possible were vivid symbols of the contemporary world. The Grillo demonstrated that the telephone was not just a useful piece of household equipment, but could also be an object of desire – something both useful and attractive at the cutting edge of technology and design.

### MARCO **ZANUSO**, RICHARD **SAPPER**

#### 1916–2001, 1932–

Marco Zanuso trained as an architect, designing many buildings both in his native Italy and abroad. After World War II, he turned to industrial design, working from the late 1950s with German designer Richard Sapper (see p.200), who had moved from Daimler Benz to work in Milan with the celebrated architect Gio Ponti. Zanuso and Sapper collaborated until 1977, producing some of the most famous designs of the 1960s and 1970s, from their child's plastic stacking chair for Kartell to electrical appliances for Brionvega. From the 1970s onwards, Zanuso continued to produce work independently for Italian clients.

▲ **Marco Zanuso (left) and Richard Sapper** (right) with the Grillo telephone, 1967

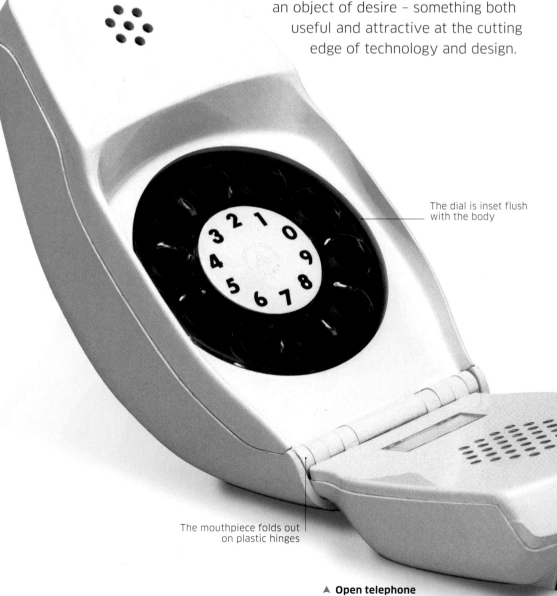

The dial is inset flush with the body

The mouthpiece folds out on plastic hinges

▲ **Open telephone**

# Visual tour

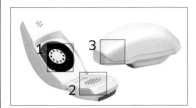

## KEY

▶ **RECESSED DIAL** To save space and enable the unit to fold, the designers recessed the dial into the body. The dial does not, however, have a traditional stop. Instead, when the user inserts a finger into one of the holes, a button activates a pin that pushes through the back of the dial. The pin stops the dial rotating once it has reached the end position.

1

2

▲ **FOLD-OUT MOUTHPIECE** The telephone flips open to reveal the mouthpiece. This bears a network of holes arranged in a neat hexagonal pattern, behind which is the microphone. This fold-out design proved very influential, not just for home telephones but, in a later, miniaturized form, as a way of making mobile phones compact.

▼ **CLAMSHELL BODY** Zanuso and Sapper took full advantage of the sculptural quality of plastic when they designed the Grillo. The curves of the outer casing give the phone a stylish appearance, and the dial unit is pleasantly smooth and feels comfortable to hold.

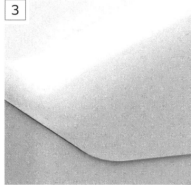

3

## ON **DESIGN**

Zanuso and Sapper's designs are famous for exploiting the sculptural properties of new materials, as well as for embracing technological advances such as the transistor, which meant that devices like radios and televisions could be made much smaller than before. Their celebrated collaborations include innovative audiovisual devices for the Italian manufacturer Brionvega, such as the Algol and Doney television ranges. In a departure from the standard TV box shape, the Doney body followed the curves of the screen, which took up the entire front. The controls were set discreetly on the top of the unit so that the viewer was not distracted. The pair put similar care and ingenuity into their TS 502 radio, also for Brionvega, with its two-part plastic body that hinges open to reveal the loudspeaker and controls. Designs like this put Zanuso and Sapper at the forefront of their field and helped to raise Milan's profile as the capital of the most innovative and stylish design.

▲ **Brionvega TS 502**, 1964

# "You have to rely on your instinct"

**RICHARD SAPPER**

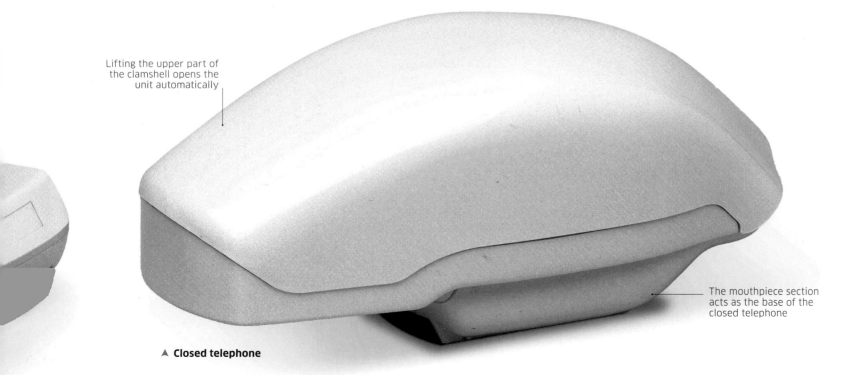

Lifting the upper part of the clamshell opens the unit automatically

The mouthpiece section acts as the base of the closed telephone

▲ **Closed telephone**

# Kodak Instamatic 33

1968 ■ PRODUCT DESIGN ■ VARIOUS MATERIALS ■ UK

SCALE

## KENNETH GRANGE

**The Kodak Instamatic cameras** of the 1960s and 1970s were some of the most successful in photographic history, selling in tens of millions all over the world. Their creator was industrial designer Kenneth Grange, and their clean lines and clearly labelled controls embodied exactly what the cameras were about – simplicity and ease of use. The Instamatic 33, with its very basic controls and uncluttered layout, was one of the best and most popular.

Like all the early Instamatics, the 33 was based around Kodak's ingenious 126 film cartridge, which was simply dropped into the camera – there was no need to thread or rewind the film. To take a photograph, you just had to set the lens for bright or overcast weather, point the camera, and press the shutter release. Both the viewfinder and shutter release are carefully placed so that your eye and finger find them automatically, while the rectangular body – designed to accommodate the film cartridge – is rounded at the corners, making the camera comfortable to hold.

Grange's design was also cheap to manufacture, partly because it did not have a complex rewind mechanism, but also because the compact camera was mostly made of plastic. The Instamatic 33 is a shining example of the way that good design can be applied to an inexpensive product, giving it mass-market appeal.

### KENNETH **GRANGE**

**1929–**

Educated at art school in London, Kenneth Grange worked as an architectural assistant before setting up his own design consultancy in 1956. He was a co-founder of the influential design group Pentagram in 1972 and has produced a huge body of work, from Japanese sewing machines to parking meters. Grange was influenced by the clean lines and analytical approach of German and Scandinavian Modernist designs, and has always made the needs of the user a priority. His designs include food mixers for Kenwood, irons for Morphy Richards, and razors for Wilkinson Sword.

# Visual tour

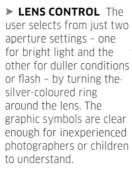

**KEY**

▶ **SHUTTER RELEASE** This takes the form of a horizontal bar parallel with the top of the camera body. When the camera is in its case, a plastic tab fills the gap beneath the button, to prevent it being accidentally pressed down, which would expose the film.

▶ **LENS CONTROL** The user selects from just two aperture settings – one for bright light and the other for duller conditions or flash – by turning the silver-coloured ring around the lens. The graphic symbols are clear enough for inexperienced photographers or children to understand.

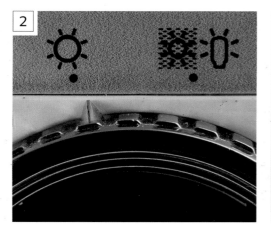

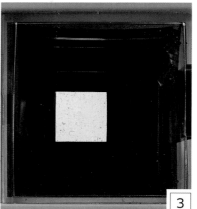

▲ **FILM ADVANCE KNOB** This is a wheel on top of the camera body. It can only be turned one way and stops moving automatically when the film has wound on to the next frame.

◀ **VIEWFINDER WINDOW** The clear, bright viewfinder is square, because that is the shape of the images in the camera's 126 film format. Framing close subjects could be tricky because the viewfinder is set to one side, but owners learnt to compensate for this by moving slightly to one side when taking photographs.

## ON **DESIGN**

Although celebrated for his user-friendly household appliances, Kenneth Grange has also been an influential designer for transport. He created a contemporary, new body for the London taxi without radically altering its distinctive shape, and designed Britain's most familiar train, the InterCity 125, which was introduced in 1976. Grange was originally hired to design the livery for this major project, but in the end he also worked with engineers to create the striking, aerodynamic exterior of the power cars at the front and rear of the train, as well as designing the interiors of the carriages. The angled profile and streamlined form of the 125 have not dated, and the trains are still in use on Britain's rail network.

▲ **InterCity 125 train**, Kenneth Grange, 1976

## IN **CONTEXT**

The British multidiscipinary design studio Pentagram was founded in London in 1972 by Theo Crosby, Alan Fletcher, Colin Forbes, Kenneth Grange, and Mervyn Kurlansky. The idea was to create a partnership of equals who could offer a range of design services including architecture, interior design, product design, corporate identity, and graphics. The studio was very successful, producing some of the most memorable work of the late 20th century, and expanding to open offices in the USA and Germany. Over the last 50 years, the group has built up a long client list of high-profile companies ranging from Nike to Tiffany, Penguin Books to Saks Fifth Avenue. Their architecture and interior work includes restaurants, shops, and banks, as well as domestic buildings, and they have done much to bring good design to both the home and the high street. Kenneth Grange retired from Pentagram in 1997, after 25 years with the partnership.

▲ **Identity for the Museum of Chinese in America**, Pentagram, 2009

The model name is in widely spaced sans serif capitals

The lens setting for bright sunshine is a simple sun symbol

When the user fitted a flash, this lens setting was used

A leatherette panel on the front makes the camera easy to grip

# Valentine typewriter

1969 ▪ PRODUCT DESIGN ▪ ABS PLASTIC AND OTHER MATERIALS ▪ ITALY

SCALE

## ETTORE SOTTSASS, PERRY A. KING

**In the late 1960s, the typewriter** was the most recognizable piece of office equipment, but many people also used them at home. The best, including those produced by the design-conscious Olivetti company in Italy, were good examples of functionalism. However, most machines were dull and uninspiring to look at. The Valentine typewriter, when it appeared, was completely different – brightly coloured and user-friendly, with an ingenious travelling case. It was a typewriter designed for use in the modern home, rather than the corporate office. However, its striking appearance, which was the work of Ettore Sottsass in collaboration with British-born designer Perry A. King, sparked a radical reappraisal of office design.

The Valentine works in the same way as every manual typewriter of its time – the keys move a series of levers to make metal type bars hit the paper through an inked ribbon. What the designers changed was the material of the body, from traditional metal to moulded plastic, and they introduced a fun range of colours – most famously, red, but also white, green, and blue. They also devised the ingenious way in which the typewriter, which has a carrying handle attached to the rear section, tucks neatly into its plastic case. This combination of functionality and brilliant colour helped to transform the way both designers and users thought about the machines they used every day in the office and at home.

The carriage-return lever folds back when not in use

The ribbon spools have bright orange-yellow caps

Broad space bar

### ETTORE **SOTTSASS**

#### 1917–2007

Born in Austria, Ettore Sottsass spent most of his life in Italy, where he studied architecture in Turin. Over a long career, he created buildings and produced original designs for more than 100 clients from his Milan practice. His work for Olivetti included typewriters, adding machines, Italy's first electronic calculator, computers, and office furniture. His designs for other clients ranged from ceramics influenced by trips to India to furniture inspired by American Pop Art. By the late 1960s, Sottsass was a leading figure in the Italian "Anti-Design" movement – whose followers embraced bright colours, ornament, and the mass media in a reaction against the pared-down aesthetics of Modernism. In 1981, Sottsass co-founded the cutting-edge design group Memphis (see p.211). When it disbanded, he continued to practise, mainly as an architect, and to create innovative, colourful, and often provocative designs.

IN **CONTEXT**

Two vibrant movements, Anti-Design and Radical Design, started in Italy in the 1960s. Both were concerned to distance themselves from Modernism and to reject the idea that design could be used as a marketing ploy without considering its social or cultural effects. Founded in Florence in 1966, Superstudio was one of the most important Radical Design groups. Its members were keen to explore new, utopian lifestyles and their work involved teaching, holding exhibitions, and producing manifestos, as well as creating striking designs, such as the Gherpe lamp and the laminate-covered Quaderna console table.

➤ **Gherpe** perspex lamp, Superstudio, 1967

The end of the carriage aligns perfectly with the body for efficient packing

A scale makes it possible to set margins and tabs

The tab key is the only red key

The body is flat and compact

Black plastic clip

The travelling case doubles as a wastepaper basket

▲ **Side view**

▲ **Case**

# Visual tour

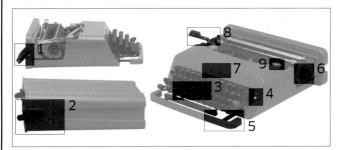

**KEY**

▶ **CARRYING HANDLE**
Attached to the back of the typewriter, the handle folds out for carrying when the machine is stowed in its case. It is strong enough to take the weight of the typewriter, and is the same angular shape with rounded corners.

1

▶ **CLIPS** These flexible plastic clips attach to the body of the typewriter to fix the machine securely in its case. The designers chose black to ensure the clips stood out, and their size and shape make them easy to use.

3

◀ **COLOUR SELECTOR**
Like most typewriters, the Valentine has a two-colour ink ribbon. A small lever to one side of the keyboard is used to select the colour. On the Valentine the colour settings are simple round buttons. In the central, white setting, however, the type bars do not strike the ribbon. Instead, they produce an embossed print-out intended for stencil-based copying systems.

4

▲ **KEYS** The black keys have a slightly concave upper surface, to make them more comfortable to strike than the flat keys often fitted to older typewriters. The letters, numbers, and symbols on the keys are in a modern style of lettering. It is slightly square and was designed to be both clear and contemporary in appearance.

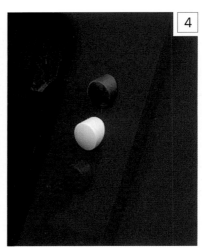

5

▶ **FRONT CORNER** A distinctive feature of the Valentine is the break in the part of the plastic framework near the front of the machine. It looks and functions almost like a second handle, and can be used to position the typewriter on the desk.

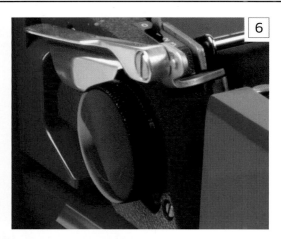

◄ **PLATEN KNOB** The knob on the end of the platen (roller) enables the user to insert the paper and move it to the right position for typing. This is a plain, round knob but it stands slightly raised from the main body.

◄ **BRANDING** The model name is emblazoned across the front of the machine in a typeface called Braggadocio. The letter forms, which are based on those used for stencilling, look bold and modern, yet the typeface was actually designed in 1930. The same typeface was used on the packaging and the user manual.

▲ **CARRIAGE RETURN LEVER**
This lever is for rolling the paper on and resetting the carriage to type the next line. To stow the typewriter in its case, the user folds this lever down and locks the carriage.

◄ **RIBBON SPOOL CAP** These orange-yellow caps, which secure the spools, can be unscrewed to fit a new ribbon. Some typewriters require you to remove the top of the body to reach the spools, but the Valentine's are far more accessible.

## IN CONTEXT

Olivetti had been employing innovative designers since the 1930s, and already had a reputation as one of the most forward-looking companies in its field by the time that Sottsass started working there. His work was later continued by Mario Bellini, whose yellow Divisumma 18 calculator, with its soft rubber keyboard, showed that the most mundane piece of office equipment could be made into something innovative, easy to use, and fun. The more sober Divisumma 28 (in grey but with a similar soft keyboard) and the wedge-shaped Logos 50/60 calculator were more examples of the flair Bellini brought to the office. By the 1990s, when personal computers were becoming more widespread, designs such as Sottsass's Philos 33 notebook and Michele De Lucchi's Echos 20 laptop helped to cement Olivett's reputation for outstandingly stylish and functional products. The company also produced office furniture, including the Ephesos range by Antonio Citterio, and dominated the field in Europe.

► **Olivetti Divisumma 18 calculator**, Mario Bellini, 1973

## ON DESIGN

Olivetti had been making typewriters since 1908, but their products were not widely recognized until after World War II. Models such as the 1948 Lexikon 80, a big, curvaceous, organic-looking machine in enamelled aluminium helped put the company on the map. The brainchild of Olivetti's chief designer, Marcello Nizzoli (see p.148), the Lexikon replaced the black, old-fashioned typewriters of the pre-war years. The company also produced smaller models, aimed at the home market, notably Nizzoli's Lettera 22 of 1949–50. This lightweight, portable typewriter, available in pale blue or grey, became very popular with journalists on the move. In 1963, it was followed by the Lettera 32 by the same designer, which sold very well. Both of these machines were compact, reliable, and came with carrying cases. The mould for the Lettera 22 and 32 was used to make the Valentine in plastic a few years later.

► **Olivetti Lettera 32**, Marcello Nizzoli, 1963

# Revolving cabinet

1970 ▪ FURNITURE ▪ PLASTIC AND STEEL ▪ JAPAN

SCALE

## SHIRO KURAMATA

**Japanese designer Shiro Kuramata** began his career designing boutique interiors at a time when the unconventional clothes of Japanese fashion designers such as Rei Kawakubo (founder of the Comme des Garçons label) were beginning to attract attention. Having developed a deep respect for the minimalist elegance of traditional Japanese interiors, Kuramata injected his own sense of subversive playfulness and created some of the most idiosyncratic and original furniture of the 1970s and 1980s.

For one of his most celebrated pieces, the Revolving cabinet, Kuramata took a radical approach to the design and dispensed with the outer carcass that usually encased a set of drawers. Instead, he arranged the glowing red drawers at precise intervals on an upright, cylindrical column, which they could pivot freely around. The stability of the design allows the drawers to be stacked vertically, to resemble a more conventional chest of drawers, or arranged much more creatively in many possible combinations. When the drawers are opened out, the bright-red cabinet resembles a piece of abstract sculpture, transforming what is usually a solid-looking piece of furniture into a light and insubstantial structure. Even though there are 20 drawers (plus the base unit), the cabinet retains a sense of harmony and balance – the qualities of the traditional Japanese architecture that inspired Kuramata. The piece brought him international recognition, proving as popular in the West as in Japan. It was later manufactured by the Italian firm Cappellini.

The corner of the drawer masks the central column

The drawers rotate through 360 degrees

The unit stands on four small black feet

### SHIRO **KURAMATA**

#### 1934-91

Trained in woodworking and interior design in Tokyo, Shiro Kuramata's first job was with a furniture manufacturer. In the 1960s, he opened his own office in Tokyo, and worked on the interiors of some 300 shops and restaurants in Japan. By 1970 he was also acquiring a reputation for his unconventional furniture. In the last 20 years of his life, he was a prolific furniture designer, his work marrying his quirky sense of humour with an innovative choice of materials, including acrylic, glass, and concrete. Combining Japanese minimalism with Western Postmodernism, Kuramata's designs were highly sought after, and he worked both with Japanese companies and Western design groups such as Memphis (see p.211).

# Visual tour

**KEY**

> **UPRIGHT** The column that supports the drawers is finished in black and is only visible in the small gaps between each drawer. As well as physically supporting the drawers it is an anchor, acting as a central spine from which the opened drawers fan out.

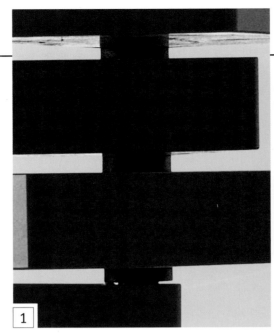

1

3

2

▲ **BASE** Because the unit is about 1.85m (6ft) tall, it needs to sit firmly on the floor if it is not to topple over. The base unit of the cabinet is therefore heavily weighted, and the weight is concealed in the bottom unit, which looks at a glance just like another drawer.

◄ **DRAWERS** The drawers are made of polished acrylic in vibrant red, showing the love of bright colours that endeared Kuramata to the design group Memphis later in his career. The drawers are quite shallow (about 6.9cm/2¾in deep), and have no lids. For many owners, their sculptural quality takes precedence over their storage capacity.

## ON **DESIGN**

Shiro Kuramata had quite a keen interest in Postmodernism, especially its use of strong colours, its sense of humour, its cultural references, and its love of reinterpreting traditional shapes and forms in new guises. Several of his pieces of furniture have names referring to movies and the theatre - the Miss Blanche chair, with artificial roses embedded in its Plexiglas form, alludes to the character Blanche Dubois in Tennessee Williams's play
*A Streetcar Named Desire*; the immaterial appearance of the chair refers to Blanche's world of dreams and illusions. The piece also references design history, because the shapes made with the Plexiglass - the gently curving arms and straight back - interpret traditional armchair design in a fresh and humorous way.

▲ **Miss Blanche armchair**, Shiro Kuramata, 1988

## IN **CONTEXT**

Shiro Kuramata's 1970 series, Furniture in Irregular Forms, displayed his radical reinterpretation of conventional shapes and helped bring him to the attention of a worldwide audience. This series included "wavy" chests of drawers that looked strangely distorted. In fact, they were beautifully crafted minimalist works: the drawers fitted perfectly, one above the other, and were easy to slide in and out of the curving body.

In Kuramata's 18-drawer cabinet, the white drawers are contained in a simple, black outer frame and the handles are plain, cylindrical pieces of metal. The piece is beautifully balanced and the proportions, like those of much traditional Japanese furniture, are harmonious. Yet there is a curious sense of movement that captures the imagination and gives the cabinet a unique, expressive quality.

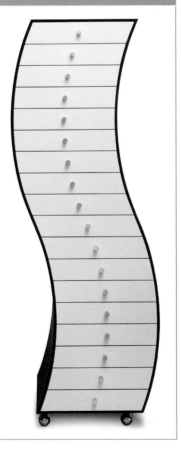

► **18-drawer cabinet** from the Furniture in Irregular Forms series, Shiro Kuramata, 1970

# Beogram 4000

1972 ▪ PRODUCT DESIGN ▪ WOOD, ALUMINIUM, STAINLESS STEEL, PLASTIC, AND RUBBER ▪ DENMARK

SCALE

## JACOB JENSEN

**In the 1960s and 1970s**, hi-fi equipment consisted of components housed in metal boxes that looked more at home in the laboratory than the living room. The electronics designer's work was to create circuits that worked, with little thought given to how products looked. Danish manufacturer Bang & Olufsen, in collaboration with industrial designer Jacob Jensen, changed this, transforming record-players for vinyl discs into technologically innovative products that were also sleek, desirable objects.

The stylish Beogram 4000 looked unique. It had a state-of-the-art turntable set flush with the surrounding casing, a plastic lid that lowered slowly, and controls that were simple to operate and virtually invisible – unlike the levers and switches on conventional turntables. The system's sound reproduction was advanced: Jensen and the engineers at Bang & Olufsen rethought the tone arm, the device that holds the stylus in contact with the record. Most tone arms swung from a pivot point, so the angle at which the stylus tracked the record changed continuously, affecting sound quality. The arm on the Beogram was set at a fixed tangent to the record and moved across it. A second arm, which contained a light source and sensor, "read" the record, detecting its size and setting the turntable to spin at the correct speed, although the user could also set the speed manually if the record was a non-standard size. There were many hidden refinements too, including a suspension system that protected the turntable from vibrations. The whole unit was an amazing combination of style and performance, and set a new standard for user-friendliness, sound reproduction, and looks.

The transparent lid drops silently to close

The slimline plinth is lined with a narrow band of teak

### JACOB **JENSEN**

#### 1926–

Jacob Jensen trained as an upholsterer and had begun to design furniture when he enrolled at the School of Applied Arts in Copenhagen in 1948. He was the first graduate of a new industrial design course devised by architect Jørn Utzon, and found work with design partnership Bernadotte & Bjørn. By the early 1960s, he had set up his own practice, been appointed Assistant Professor of Industrial Design at the University of Chicago, and become a partner in a New York design office. He maintained his practice in Denmark, however, and by the 1970s, his work for Bang & Olufsen was becoming well known. Jensen also designed office chairs for Labofa, toys for LEGO, kitchen appliances for Gaggenau, and graphics for various clients. Jacob Jensen Design is a family enterprise.

## ON DESIGN

Jacob Jensen worked with Bang & Olufsen between 1965 and 1991 and was largely responsible for the combination of stunning, minimalist style and high-quality products that made the company's reputation. He produced a range of hi-fi equipment from loudspeakers to sound systems, and also the Beolit 600 and 400 portable radios. Jensen's other designs, which included office furniture, kitchen appliances, and even a wind turbine, all display a similar sophistication.

➤ **Beovox 2500 Cube speaker**, 1967

# Visual tour

**KEY**

1

▲ **TONE ARM** Because it moves at a tangent to the disc, the tone arm (on the right) could be made short and stiff. This helped the precision tracking, accurate to within a fraction of a millimetre (1/32 of an inch), and enhanced sound reproduction.

◄ **TURNTABLE** The turntable's platter is cast in heavy aluminium. It has radial rubber bars that are raised slightly and designed to support both 7-inch and 12-inch records.

2

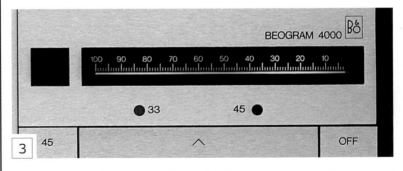

3

▲ **DISPLAY** This illuminated scale indicates the precise speed of the turntable. Fine adjustments could be made with a screwdriver to the sophisticated electronics, for example, to correct recordings that were at slightly the wrong pitch.

➤ **LID** The large lid, which covers the whole unit, is made of clear plastic with a smoke-coloured tint. It is spring-loaded and closes slowly and gently, so that it does not jolt records while they are playing.

4

# Tizio desk lamp

1972 ■ LIGHTING ■ ALUMINIUM AND ABS PLASTIC ■ GERMANY/ITALY

SCALE

## RICHARD SAPPER

**When German designer Richard Sapper** was looking for a new work light, he could not find one that met his requirements. Sapper's solution was to design his own lamp, a low-voltage model for the Italian company Artemide. Sapper's sculptural and technically innovative Tizio was a radical reinterpretation of the desk lamp. It won several international awards and soon became a design classic.

At the heart of the Tizio are two pairs of aluminium arms. Instead of being held in position by springs, as on earlier desk lamps such as the Anglepoise (see p.72), these arms are counterbalanced with weights. This system, together with the rotating base, means that the lamp head can be moved into a range of positions with very little effort. Another innovation is that the electric current travels from the transformer (hidden in the base) to the bulb directly through the metal of the aluminium arms. The lack of wires in the arms has several benefits – the lamp is easy to assemble, and, although the arms are very thin, they are perfectly balanced because there are none of the variables that come with slight differences in the thickness and weight of the wire. In the head of the lamp is a halogen bulb, a type of bulb previously used mainly in the car industry for its bright, concentrated light. The small size of the bulb was also key in meeting Sapper's desire for a compact head. All these carefully considered features make the Tizio lamp a precise marriage of form and function. It is immaculately finished, the bird-like head is beautifully angled, the counterweights have gently curved bases, and the whole lamp is an elegant matt black with a few eye-catching red highlights. No wonder that possessing a Tizio lamp came to symbolize an appreciation of the very best in modern design.

The arms conduct low-voltage electricity to the lamp head

The lower counterweight can pass through the upright supports, enabling the arms to be set vertically

---

### RICHARD **SAPPER**

**1932–**

After studying philosophy, graphic design, and engineering in Munich, Richard Sapper worked for Daimler Benz. He moved to Italy, where he worked in the studio of Gio Ponti and Alberto Rosselli, then collaborated with the Italian designer Marco Zanuso (see p.188) on a wide range of innovative products, including radio and TV sets for Brionvega. Since the 1970s, Sapper has carried out consultancy work for Fiat, Pirelli, and IBM, and has designed appliances for Alessi that blend high-tech and postmodern approaches. In Sapper's strong, functional creations, precision and attention to detail combine with outstanding design flair.

# "I wanted a small head and long arms... And I wanted to be able to move it easily"

**RICHARD SAPPER**, on his ideal work lamp

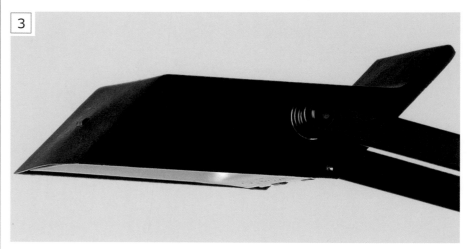

The arms are covered in a plastic reinforced with fibreglass that is light, tough, and an electrical insulator

## IN **CONTEXT**

Artemide was founded in Milan in 1960 by Ernesto Gismondi and Sergio Mazza, who commissioned many top designers and architects to work on their range of high-quality lighting products. The company has marketed designs by prominent architects such as Norman Foster and Mario Botta, and by influential artists and designers including Ettore Sottsass (see p.192) and Luigi Serafini. The Tizio is their most famous design, closely followed by another adjustable lamp, the celebrated 1987 Tolomeo, designed by Michele De Lucchi and Giancarlo Fassina. The Tolomeo takes the opposite approach to Sapper's lamp, with hidden springs and steel tension cables to keep the adjustable head in position.

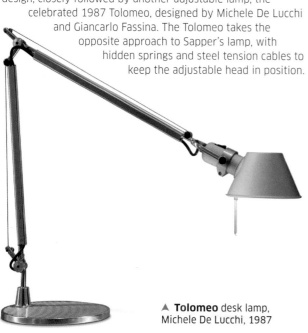

▲ **Tolomeo** desk lamp, Michele De Lucchi, 1987

# Visual tour

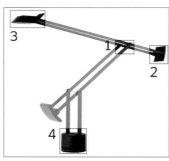

**KEY**

1

▲ **PIVOT POINTS** Where the arms pivot, they are joined by chrome-plated bars with red, moulded-plastic insulators. The main joints in the lamp are connected with snap fasteners, so that if the lamp is dropped they come apart rather than breaking and can easily be fitted together again.

◄ **COUNTERWEIGHTS** The two metal counterweights balance the arms perfectly – they move at the touch of a finger. The weights are the same width as the pair of arms so that they can swing inside the wider supports below, giving the arms a long reach when they are fully extended.

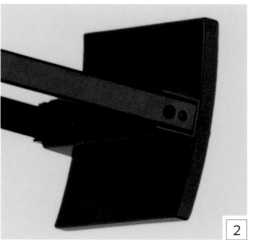

2

3

▲ **LAMP HEAD** Halogen bulbs heat up when in use, so the lamp housing has ventilation holes. The user can adjust the head safely using the arms, or can grasp the protruding part of it. The interior of the head is shaped to direct the light precisely and evenly.

◄ **BASE** The solid base swivels on its disc, so that the lamp can be turned 360 degrees, while the weight of the transformer inside helps to keep the lamp stable. Openings across the top and around the bottom of the base ventilate the transformer, and the red on/off switch has settings for two different intensities of light.

4

# Wiggle chair

1972 ■ FURNITURE ■ CORRUGATED CARDBOARD AND HARDBOARD ■ USA

SCALE

## FRANK GEHRY

**In the late 1960s, Canadian architect** Frank Gehry began to experiment with cardboard furniture. He was already familiar with the qualities of corrugated cardboard, which he had used to make architectural models, and he now discovered that if he glued sheets together in layers, with the corrugations running in alternate directions (like the layers of wood in plywood), he could create a material that was very strong. Gehry had overcome the strength problems characteristic of cardboard furniture, and come up with a robust new material, which he called edge board.

By bending the layered, corrugated cardboard and sculpting it with a hand saw and a knife, Gehry created graceful, curvilinear shapes, which he developed into his first line of furniture, the Easy Edges series. One of the most original pieces was the Wiggle chair, which had a slightly sloping back extending from a base arranged in a series of loops. The looped base was stable, the back provided good support, and the whole chair was very strong. The surface of the board felt soft, but also hardwearing. Only the edges, where the flat surfaces of the cardboard were exposed, were weaker, so Gehry strengthened them with hardboard laminate.

The unconventional Wiggle chair and the rest of the Easy Edges range (which included more chairs, a stool, tables, and a bed frame) attracted a great deal of attention when they were launched in American stores. The low cost of the materials also meant the furniture was reasonably priced, but Gehry was keen to pursue his career as an architect and did not market the furniture commercially. As a result, early Wiggle chairs, which once sold for $35 each, are now expensive collector's items.

## "Creativity is about play and a kind of willingness to go with your intuition"

**FRANK GEHRY**

# Visual tour

**KEY**

▶ **SURFACE** The rippled surface of the corrugations is visible on the chair back. The velvety texture of edge board has been compared to suede or corduroy. It is strong enough unvarnished, as on the Wiggle Chair, for everyday use.

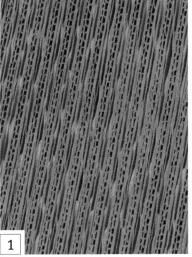

1

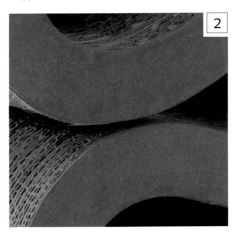

2

◀ **CHAIR EDGE** The smooth side of the chair, which is covered in painted hardboard, forms a contrast to the more rugged appearance of the seat and back. The sinuous curves give the chair an organic, sculptural effect that is typical of Gehry's designs, defying the convention that furniture (or buildings) should be composed of straight sides.

## FRANK **GEHRY**

### 1929–

After training in architecture in Los Angeles and at Harvard, Canadian-born Frank Gehry set up his own practice in California in 1962. One of the world's most sought-after architects, he designs buildings that explore architectural form in inventive ways. Some have relied on sculptural elements, such as a vast fish at the entrance of a Japanese fish restaurant, while others, such as his famous titanium-clad Guggenheim Museum in Bilbao, Spain, enclose their spaces in a series of unpredictable but beautifully managed curves. Gehry designed another collection of cardboard furniture, Experimental Edges, and chairs for Knoll made of interwoven plywood strips.

## ON **ARCHITECTURE**

Gehry's breakthrough building was the house he designed for himself in Santa Monica, California, in 1978-79. In this carefully planned building, he used industrial materials not often found in domestic architecture – corrugated metal, steel poles, and wire mesh –juxtaposing them at odd angles to create an impression of randomness. This house anticipated the ideas of Deconstructivism, a 1980s movement that treated architecture as a form of expression for ideas such as dislocation, disruption, and distortion. Gehry's style evolved in different directions later in his career, but he continued to use unusual materials to explore architectural form.

▲ **Gehry House**, Santa Monica, 1978-79

The texture of the surface is modulated by the curves

The seat slopes back slightly for comfort

The undulating loops of edge board gave the chair its name

The end of the length of edge board makes a firm base at the back of the chair

# Munich Olympic Games pictograms

1972 ▪ GRAPHICS ▪ VARIOUS MATERIALS ▪ GERMANY

## OTL AICHER

**Graphic designers often face the challenge** of presenting information pictorially, and this task is most difficult in an international context, where there can be no reliance on words. One of the most successful systems of visual communication was the series of symbols produced by German designer Otl Aicher for the 1972 Munich Olympic Games. Aicher, who was Head of Communications at the the Ulm School of Design in Germany (see p.156), had already evolved a rigorous graphic style using simple geometric forms and clear sans serif type. For the Munich symbols, he created a series of stylized figures worked out on a grid that allowed only verticals, horizontals, and diagonals. The figures, in white against a plain background, had circles for heads, lines with curved ends for limbs, and, where necessary, sports equipment depicted in a finer line. Each figure was set in a square with a narrow white border. The combination of the figure's pose and equipment made each sport simple to identify, and people quickly took to the symbols. Aicher's pictograms have become object lessons for designers of signs, charts, diagrams, and any other visual material where text is at a minimum and immediate understanding is vital.

### OTL **AICHER**

#### 1922-91

After briefly studying sculpture in Munich, German industrial and graphic designer Otl Aicher set up a studio in his home town, Ulm, in 1948. Soon afterwards he became one of the founders of the Ulm School of Design, where he taught during the 1950s and 1960s. Developing its curriculum along Bauhaus lines, the Hochschule became one of Germany's most prestigious schools of design. Aicher designed corporate identities for companies including Braun (where he also collaborated on product design), Dresdner Bank, Frankfurt airport, and Lufthansa. He also published widely on design and drawing.

## Visual tour

**KEY**

1

2

3

▲ **FOOTBALL** The football symbol is a good example of the use of diagonals to imply movement and speed. A simple circular outline the same size as the player's head is enough to portray the ball, and the truncated, angled leg suggests a kick.

◀ **CANOEING** Instead of drawing the canoe, Aicher realised it would be simpler to depict the paddle in outline. A horizontal line suggests the side of the boat and the bottom of the frame represents water.

◀ **BOXING** By adding circles to the usual curve-ended limbs, Aicher found a simple way to represent the boxer's gloves, making the figure instantly identifiable. The diagonal line of the arm symbolizes the boxer's dynamic uppercut.

## ON **DESIGN**

Pictograms were first used at the London Olympic Games of 1948 to help spectators and participants find venues and events, for publicity material, and to create a visual identity. With their heraldic, shield-shaped frames and line drawings, these pictograms were very traditional in style. Symbols were next used in Tokyo in 1964, and similar stylized pictograms have been used at all the Olympics since. Among the most successful were those for the 1968 Mexico City games, which featured graphic representations of equipment against coloured backgrounds.

▲ **London** 1948

▲ **Mexico** 1968

## IN **CONTEXT**

When he created his Munich pictograms, Otl Aicher was influenced by the work of the Austrian sociologist Otto Neurath, who developed the Isotype system of symbols with graphic designer Gerd Arntz in the 1920s. The Isotypes were designed to convey information visually, without using text. Some 4,000 simple symbols were created to display all kinds of information, and were particularly suitable for representing statistics – indicating numbers by using more or fewer pictograms. The symbols were designed to be easily understood by people who could not read and across language barriers. They were applied to a range of fields, from industrial production to religion.

**Industry**

**Population**

**Food production**

**Politics**

# "In design man becomes what he is. Animals have language and perception as well, but they do not design"

**OTL AICHER**

# Suomi tableware

1976 ▪ HOMEWARE ▪ PORCELAIN AND METAL ▪ FINLAND/GERMANY

## TIMO SARPANEVA

**The Suomi (Finland) porcelain service**, created for the German company Rosenthal, displays the sculptural shapes and characteristic attention to detail of Timo Sarpaneva, its renowned Finnish designer. Drawing inspiration from pebbles that have been gradually eroded into rounded shapes and polished by water, Sarpaneva devised a form that combined the circle within an outline consisting of four curving sides and four rounded corners – a shape that led some writers to say that the designer had succeeded in "squaring the circle". To arrive at the finished design, Sarpaneva spent several years making wooden models – trying out the various pieces in the service until he had achieved the pure forms he sought. At first, Rosenthal thought the design, which was produced in plain white, too simple, but their doubts proved to be unfounded:

Suomi combines the precision of industrial design with organic forms that owe much to the Finnish craft tradition. The service is elegant, but also practical: the pieces are comfortable to hold, and the details – from the rim that keeps food on the plate to the broad handles on the cups – all work well. In 1976, the design won the Italian Faenza Gold Medal. Rosenthal has kept the service in production, and it is one of the company's most successful studio lines. In addition to the classic white version, more than 150 artists have been invited to decorate it.

### TIMO **SARPANEVA**

#### 1926–2006

After attending art school in Helsinki, Finnish designer Timo Sarpaneva went to work for the celebrated glass company littala. He was engaged in both product and exhibition design and created the company's logo (the distinctive "i" in a red circle is still used), as well as developing new techniques for mould-blown glass. Sarpaneva created glassware for other leading companies, including Venini in Italy and Corning in the United States, and produced successful textile designs in the 1960s and 1970s. He was a multi-talented artist with a strong sensitivity for materials, and he always liked to be closely involved in the manufacturing processes of his designs.

The stainless steel handle takes the form of a contrasting strap

"Shoulders" accommodate the handle joints

The simple, straight spout pours well

▲ **Coffee pot**

# Visual tour

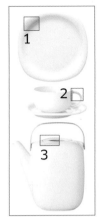

**KEY**

**▶ RIM OF PLATE** The plates have a circular inner section surrounded by a curved, four-sided edge, so the rim is uneven in width, broadening out at the corners. This makes the plate easy to pick up and gives it an unusual geometric form.

**▶ CUP HANDLE** The handles in the Suomi range are formed from a broad strip of porcelain that is very narrow in profile. The generous width makes them easy and comfortable to hold, while the slender outline, which echoes the thin walls of the cups, adds to their elegance.

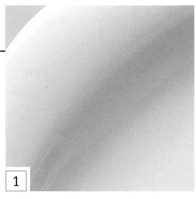

1

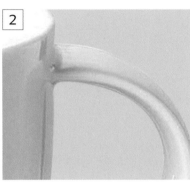

2

**◀ COFFEE-POT LID** The lid is recessed into the top of the body and has a deep hollow in the centre. The lid handle (a narrow band of porcelain matching that of the cup) is a shallow convex curve, which completes the clean line of the coffee pot.

3

## ON DESIGN

Timo Sarpaneva was a versatile designer who worked with textiles, wood, metal, and ceramics, but he is best known as a glass designer. Among his most innovative glassware was the Finlandia range, which was created using moulds made of alder wood. Glassblowers usually discarded the moulds, but Sarpaneva reused them, allowing the hot glass to stay in the mould long enough for its surface to burn and randomly determine the pattern of the glass. Every time the mould was re-used, its carbonized surface was further charred, so each new piece had a subtly different, unique surface texture.

**▲ Finlandia bark-pattern vase**, Timo Sarpaneva, 1964

The plate has a generous rim of variable width

The saucer has a traditional circular well

▲ **Plate**          ▲ **Teacup**          ▲ **Saucer**

■ **Carlton bookcase** Ettore Sottsass, Memphis Group

■ **The Face magazine** Neville Brody

■ **Whistling Bird kettle** Michael Graves

■ **Wood chair** Marc Newson

■ **Carna wheelchair** Kazuo Kawasaki

■ **Bookworm bookshelf** Ron Arad

■ **85 Lamps chandelier** Rody Graumans, Droog Design

■ **Vermelha chair** Fernando and Humberto Campana

■ **Dyson DC01 vacuum cleaner** James Dyson

■ **Verdana typeface** Matthew Carter

1980–1995

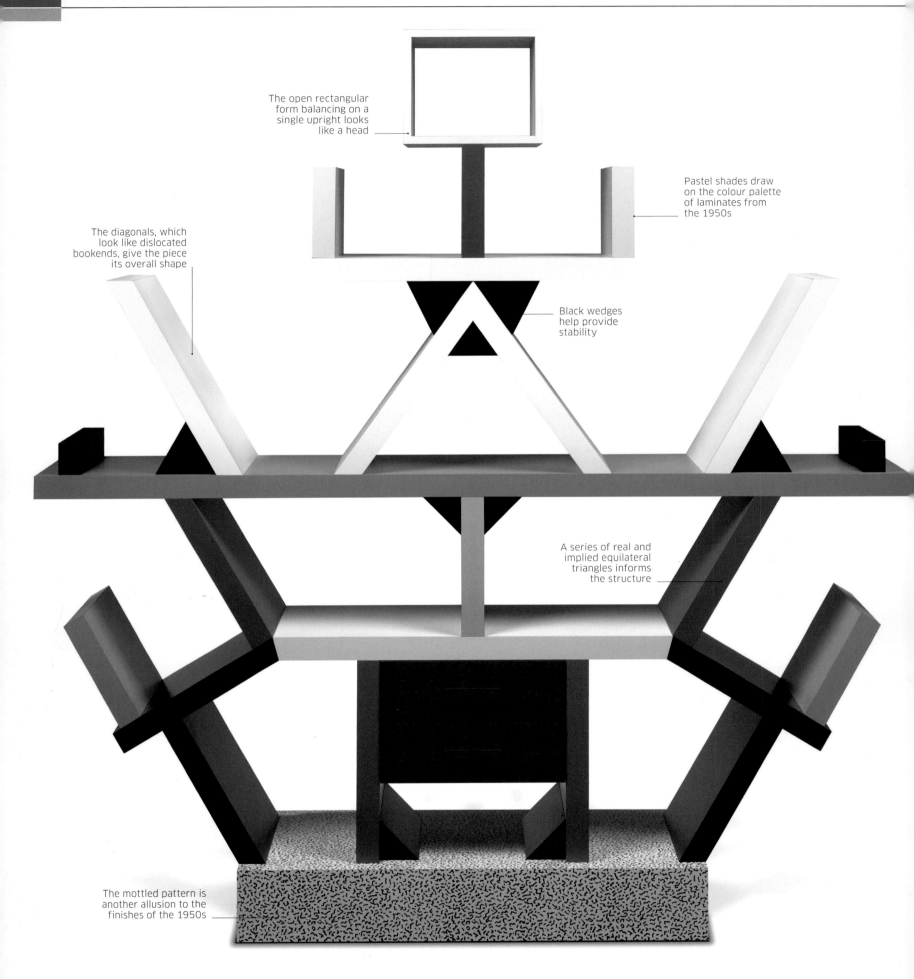

The open rectangular form balancing on a single upright looks like a head

Pastel shades draw on the colour palette of laminates from the 1950s

The diagonals, which look like dislocated bookends, give the piece its overall shape

Black wedges help provide stability

A series of real and implied equilateral triangles informs the structure

The mottled pattern is another allusion to the finishes of the 1950s

"Memphis is not new, Memphis is everywhere"
**ETTORE SOTTSASS**

# Carlton bookcase

1981 ▪ FURNITURE ▪ WOOD AND PLASTIC LAMINATE ▪ ITALY

SCALE

## ETTORE SOTTSASS, MEMPHIS GROUP

**The Carlton bookcase is a riot** of bright colours, dynamic angles, and unexpected shapes. It is one of the signature pieces of the Memphis group, a design collective formed by Ettore Sottsass (see p.192) to overturn the conventions of Modernism. For much of the 20th century, the best design had been tasteful, rational, and restrained, but the Memphis group challenged this and embraced colour, theatricality, and playfulness. The Carlton bookcase bears only a passing resemblance to a traditional piece of furniture and its shape has been likened to that of a cartoon insect. It shows how Sottsass, like other Postmodernists (see p.214), drew heavily on popular culture: with its bold colour and odd profile, the piece seems to owe something to the aesthetics of Las Vegas, reinforced by the use of inexpensive timber and plastic laminates, rather than the costly woods and finishes of most designer furniture. However, the bookcase, like much Postmodern design, embodies paradoxes. Although it is inspired by the trashy, kitsch neon and plastic of Pop culture, the bookcase is beautifully made and was intended for the luxury market. Its design may seem irrational, yet its structure, based largely on equilateral triangles, is cleverly thought out. Rarely used for storing books, it is above all a stunning piece of sculpture, widely admired and prized by collectors.

### MEMPHIS GROUP

#### ACTIVE 1981-88

The hugely influential Memphis group came together in December 1980. They took their name from the Bob Dylan song, "Stuck Inside of Mobile with the Memphis Blues Again", and the idea that Memphis was a city in ancient Egypt, as well as a centre of American blues and soul music. In the first Memphis exhibition in 1981, the furniture, ceramics, glass, textiles, and lighting designs combined cheap and luxury materials in striking combinations, with bright colours and strong patterns. Several high-profile designers were associated with the group before it disbanded in the late 1980s.

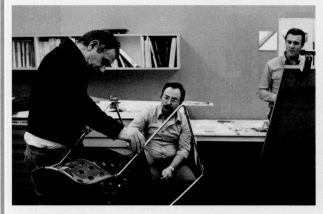

▲ **Ettore Sottsass** (left) and colleagues, Milan, 1979

# Visual tour

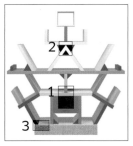

**KEY**

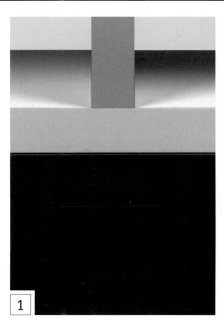

▶ **DRAWER FRONT** The unit's two drawers are very conventional in form: they are plain rectangles with simple, unassuming wooden handles. The form follows the Modernist functional aesthetic, but the red primary colour is unrestrained.

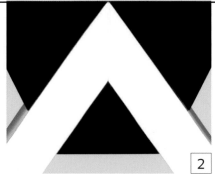

◀ **TRIANGULAR SUPPORT** The design features a number of black triangular elements, including several near the centre of the bookcase. These pieces act as supports and reinforcements, strengthening the structure. The black finish contrasts with the bold colours of most of the shelves, and throws them into relief.

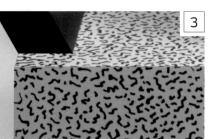

◀ **BOTTOM CORNER JOINT** One of the ingenious aspects of the Carlton bookcase is the way in which the pieces join together seamlessly, without any obvious linkages. The invisible fixings make the piece's surprising angles and strong juxtapositions of colour and pattern seem effortless.

# The Face magazine

1981-86 ▪ GRAPHICS ▪ PRINT ON PAPER ▪ UK

SCALE

## NEVILLE BRODY

**The Face** **was launched in 1980**, and the magazine soon established itself at the cutting edge of British fashion, culture, and music journalism. Its success was largely due to its unconventional use of typography and innovative layout. The magazine's art director, British graphic designer Neville Brody, focused on typography, using prominent hand-drawn type, often with tall, angular letter forms, to catch the reader's attention. The type made a huge visual impact without detracting from the strong cover photography, and there were often arresting combinations of bold and regular fonts, wide and close spacing, and tall and short letters. Brody incorporated letters and symbols into the design, took them to the edge of the page, and occasionally used type even larger than the magazine's distinctive masthead. The result was a graphic style that was strikingly different, but still clear and legible – ideal, in fact, for a magazine that was stylish, up-to-date, and constantly adapting. *The Face* captured the spirit of the times and its visual creativity under Brody set it apart from the competition. His work encouraged designers to break free from the established ways of working with typography, transforming the look of magazines, posters, and other graphics in the process.

### NEVILLE **BRODY**

**1957–**

Neville Brody trained at Hornsey College of Art, and the London College of Printing. He designed album covers before achieving recognition for his work as Art Director of *The Face* from 1981 to 1986. Other notable projects included design for *Arena* magazine and album covers for Cabaret Voltaire. Brody, a talented typographer, has also designed several typefaces for FontWorks, which he co-founded. His own company, Research Studios, has offices worldwide, and he is Dean of the School of Communication at London's Royal College of Art.

# Visual tour

**KEY**

▶ **EXTRA-TALL TYPE**
Often reproduced in colour, tall letter forms frequently featured in the magazine. As well as giving the covers a distinctive look, they enabled the designer to position large type in the narrow areas around the main photograph.

▶ **MASTHEAD** *The Face* masthead took several forms, but most featured this distinctive triangle. The bold, angular letters and wide spacing, which recall the Constructivist-inspired graphics of Rodchenko and Stepanova (see pp.44–45), made the name instantly recognizable.

2

1

4

3

▲ **PLUS SIGN** Hand-drawn symbols, such as plus signs and stars, often in bright colours, appeared on many of the covers. Here, the plus sign is used instead of a more conventional ampersand, to highlight the magazine's main features.

◀ **COLOURED PANEL** Unusual colours, type running vertically, and tinted panels were used sparingly – and usually confined to one or two words – to make a name or heading stand out from the rest of the cover.

# THE FACE

THE FACE No.62
JUNE 1985
85p ● US $2.75
GERMANY 5.80 DM

Style contenders

# WELLER
# +CONRAN

defend their titles

Vivienne Westwood
on the Collections

BULLY BEEF!

LONSDALE

+ Jesus + Mary Chain
Harry Dean Stanton
Heroin ● Miners
Bazooka ● Galliano
Japanese trash:
the best rubbish
in the world

Photo Jamie Morgan/Ray Petri

# Whistling Bird kettle

1985 ▪ PRODUCT DESIGN ▪ STEEL AND PLASTIC ▪ USA/ITALY

SCALE

## MICHAEL GRAVES

**When the Italian manufacturer Alessi** commissioned the architect Michael Graves to design a new kettle, the company stipulated two requirements. They wanted a hob kettle that would come to the boil quickly, and that looked American. Graves realized that a fast boil could be achieved by making the kettle the right shape, with a broad base that tapered towards the top, but making it look American was somewhat more of a challenge. In the end, he looked back to the childhood summers he spent on a farm in Indiana for inspiration. He remembered waking to the sound of water boiling and the rooster crowing, so he gave the kettle a whistle in the form of a brightly coloured bird that screeched when steam from the boiling water rushed up the spout.

This witty referencing of the past was typical of the Postmodern movement in architecture and design that flourished in the 1970s and 1980s. Instead of letting function dictate form, as the great Modernists had done, Postmodern designers borrowed elements of earlier styles and popular culture, and used bright, zingy colours, visual jokes, and dramatic changes of scale to challenge people's assumptions about what things should look like. They wanted to bring personality back into design and cause amusement – skyscrapers shaped like 18th-century chair backs and coffee pots that looked like buildings were among the results. The Whistling Bird kettle, with its noisy, red bird and blue handle, was part of this trend. Alessi's bestselling product to date, it has become one of the most popular and widely recognized designs of the 1980s – an icon of Postmodernism in the kitchens of millions of people all around the world.

### MICHAEL **GRAVES**

**1934–**

Born in Minneapolis, Michael Graves studied architecture in Cincinnati, Ohio, and at Harvard. He continued postgraduate studies in Rome, where the art and architecture proved a major influence. Graves both teaches and practises architecture in the United States, where, in the 1970s, he became one of a group of architects promoting Postmodernism. His Public Services Building in Portland, Oregon, was one of the movement's first internationally famous buildings. In addition to his architectural work, Graves became a prolific product designer in the 1980s, enjoying a successful collaboration with Alessi, and creating furniture, ceramics, and carpet designs. He even opened his own retail outlet. In recognition of his extensive body of work, Graves was awarded the prestigious Driehaus Prize in 2012.

The handle is moulded with indents, making it easy to grip comfortably

The generous, round knob is easy to lift for removing the lid

The integrated lid continues the subtle curve of the body

# Visual tour

**KEY**

▶ **BIRD** The red plastic bird whistles when the kettle boils, recalling children's bird-whistle toys as well as Graves's childhood memories of the rooster. The bird has to be removed to pour the water.

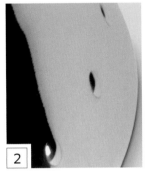

2

▲ **BASE** The shiny steel surface of the kettle is marked around the base with a row of small raised dots – a feature that is typical of the way in which the Postmodernists reintroduced ornament back into design.

3

▲ **HANDLE AND JOINT** The metal handle is joined to the body with a discreet round linkage that would not look out of place on a sober Modernist design. A red, beadlike, plastic sphere, which marks the transition to the bright blue protective grip, adds a playful touch. The strong colours catch the eye and the matte plastic surfaces provide a striking visual contrast with the gleaming stainless steel.

## ON **DESIGN**

In the early 1980s, Alessi invited some leading architects and designers, including Michael Graves, to take part in a project called the Tea and Coffee Piazza. The brief was to produce a series of silver tea and coffee sets for cultural institutions. Graves's design was architectural in form – the pieces were miniature Postmodern buildings, complete with turrets and pointed roofs. The project marked the beginning of Graves's association with Alessi, which soon led to the Bird kettle, the Euclid thermos jug, and other designs.

▲ **Tea and Coffee Piazza set**, Michael Graves, 1983

# Wood chair

1988 ■ FURNITURE ■ WOOD ■ AUSTRALIA

## MARC NEWSON

**The Wood chair uses bentwood components** in an innovative way – not to produce a lightweight, traditional chair frame, as Thonet had done with the first bentwood chair (see pp.14–15), but to create a flowing, organic form. It consists of 24 parallel strips of wood, carefully aligned with spaces between them, then bent to form a sculptural loop and two straight ends. The loop forms the seat of the chair and the sloping upper section curves inwards to form the back, while the convex lower section, reinforced by a stretcher, stabilizes the base. Five horizontal lengths of timber hold the strips of wood together.

This deceptively simple design proved hard to produce – many manufacturers said the curves were impossible to achieve – but Newson found a craftsman in Tasmania who succeeded in fabricating them from Tasmanian pine. The designer exhibited the chair in an international travelling exhibition, House of Fiction, organized by the New South Wales Crafts Council in 1988. The graceful, swelling form of the chair, similar to the organic shapes of other pieces by Newson, impressed critics, and the gently flexing strips of wood made it comfortable to sit on. Four years later, the Italian company Cappellini adopted the chair for production, helping to raise the European profile of Newson and giving the bentwood chair a new lease of life.

The chair back is slightly curved

The bentwood flexes like a sprung cushion, making the chair comfortable to sit on

## MARC **NEWSON**

### 1963–

Marc Newson went to art college in his native Sydney, where he studied jewellery design and began to make furniture using the metalwork techniques he learnt there. In the late 1980s, he moved to Tokyo, where he established a reputation for creating innovative, organically shaped furniture using materials ranging from aluminium to felt. Since then, Newson has diversified, designing products from watches to furniture and cars, and receiving numerous international awards. He is also well known for his work in the aviation industry, including the interior design of airliners for Qantas.

## ON DESIGN

Marc Newson is famous for his chairs, many of which, like the Embryo chair below, are curvaceous and biomorphic in form. The Embryo chair is striking not only for its bulbous shape, but also for the fabric in which it is covered. Newson, a keen surfer, was disappointed in the choice of conventional upholstery fabrics available for the Embryo, so he decided to try neoprene, the material used for wetsuits, instead. Available in bright colours such as green and yellow, as well as black, the neoprene gives the chair a distinctive feel and character.

◄ **Embryo chair**, Marc Newson for Idée, Japan, 1988

# Visual tour

**KEY**

1

➤ **TOP EDGE** The horizontal strip of timber at the top of the chair back holds the strips of bentwood together and provides stability. The strips of wood are secured in place with 24 metal screws.

Production models of the chair are made of natural beech; the original model was of Tasmanian pine

The seat tilts gently downwards at the sides

The bottom stretcher has a straight edge, to provide a firm base

2

▲ **SEAT AND BACK** The chair is a study in the interplay of curves and straight edges – the ribbon-like loop of the seat forms a dramatic contrast to the sloping back and rear support. The way in which the narrow strips of wood – each one a slightly different length from those next to it – interlock at this point demands great precision in manufacture.

## IN CONTEXT

Although Marc Newson's furniture has often been produced in small runs, he has also designed mass-produced items, including homeware for Alessi and lighting for Flos.

Watches have formed an important part of Newson's work from very early in his career. He designed his first Pod wristwatch in 1986, and other versions, in addition to Pod clocks, have appeared since. The watches are all minimalist in style, with plain polished metal bezels and faces, usually without numerals. As in much of Newson's work, simple, curvaceous form is allied with minute attention to detail and finish.

➤ **Pod watch**, Marc Newson, 1986

# Carna wheelchair

1989 ■ WHEELCHAIR ■ TITANIUM, ALUMINIUM HONEYCOMB, AND RUBBER ■ JAPAN

SCALE

## KAZUO KAWASAKI

**In 1977, Japanese designer Kazuo Kawasaki** was injured in a car accident, and both his legs were paralysed. This event changed both his life and the direction of his work. He subsequently designed the Carna folding wheelchair, named after a protective Roman goddess of health and wellbeing. Kawasaki devised the chair to meet his own needs in the hope that it would also be used by others. He aimed to create a practical, adjustable, and stylish chair, in contrast to the dull, utilitarian products on offer for wheelchair users.

Lightness was a priority for Kawasaki, so for the frame he specified titanium – the lightweight metal widely used in the aerospace industry and in sports equipment such as racing bicycles. Another strong and light aerospace material, aluminium honeycomb – metal pressed into a hexagonal grid and sandwiched between two flat sheets – was used in the large wheels. These advanced materials enabled Kawasaki to create a chair that weighed just 6kg (13lbs) and was easy to manoeuvre. With a removable, folding seat and back, the whole wheelchair collapsed down to a compact size, making it both portable and easy to store away. From the stunning seat with its cushioned rubber pads to the bright red finish of the wheels, the wheelchair was both practical and pleasing to look at. It was also – as Kawasaki had originally hoped – just as striking as the latest pair of trainers.

### KAZUO **KAWASAKI**

#### 1949–

Kazuo Kawasaki studied at Kanazawa College of Art, graduating in industrial design, then working in the audio products division of Toshiba. Since the late 1970s, he has worked independently, designing a range of products including spectacles with interchangeable parts, a clock, and an acclaimed rotating flat-screen computer monitor. Kawasaki has been a consultant to Apple and set up the successful Takefu Knife Village, where traditional craft methods are used to produce knives in contemporary designs. He also promotes humanitarian design and has produced a vaccination kit that can be dropped by plane into inaccessible areas. An eminent and respected teacher of design, Kawasaki has exhibited his work widely, and he has won numerous international awards.

## Visual tour

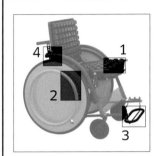

KEY

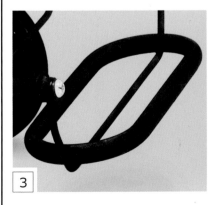

1

▲ **SEAT** The black rubber seat is designed to be both comfortable and easy to squeeze into a small space when the chair is folded away. The egg carton-like material is also functionally important as it reduces sores from sitting for extended periods of time.

2

◄ **WHEEL** The Carna wheelchair has large rear wheels that make the chair easy to propel and provide a very smooth ride. The bright red side discs conceal the tough but lightweight inner metal core.

3

▲ **FOOTREST** The simple design of the compact metal footrest is functional, keeping the overall weight of the chair to a minimum. It matches the curved frame of the chair in shape. Such careful attention to detail enhances the integrity of the overall design.

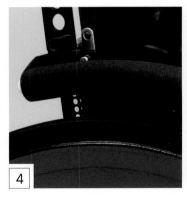

4

▲ **ARMREST** The padded armrests form an important part of the chair. Their curves follow the edges of the back wheels before they straighten to link with the small front wheels. The breadth of the armrests protects the user from contact with the rotating wheel.

# "To be a visionary designer I want to design products for myself first"

**KAZUO KAWASAKI**, 1977

During the 1980s, when Kawasaki was developing his wheelchair, Japanese design was changing rapidly. Although many designers were employed by large corporations, some pursued different paths: Shiro Kuramata (see p.196) designed furniture made of plexiglas or steel mesh; some promoted the idea of *kawaii* (cute) products, such as loudspeakers in the shape of bubbles; and others turned to traditional crafts. This variety of objects and forms, and the trend towards unusual materials fostered by Kuramata, continued after the Japanese economy declined in the 1990s, when the power of the big companies dwindled and more designers took up a freelance career.

▶ **How High the Moon chair**, Shiro Kuramata, 1986

The handle is made of lightweight metal

The material of the armrest mirrors the colour and texture of the tyres

The deep seat is made of soft, flexible rubber

Occasional touches of yellow contrast with the red wheels and black frame

# Bookworm bookshelf

1993 ▪ FURNITURE ▪ SPRUNG STEEL OR INJECTION-MOULDED PVC ▪ UK

SCALE

## RON ARAD

**At the Milan Furniture Fair of 1993**, British-based designer and architect Ron Arad unveiled a new bookcase. Instead of the conventional rectilinear design, Arad's Bookworm was a length of sprung steel with eleven brackets that held the books in place. The Bookworm had no set shape – its user could unwind it and mount it on the wall, manipulating its shape by moving the brackets – and its curving but adaptable form represented a new way of thinking about furniture. The design caught the eye of Italian furniture-maker Kartell, who agreed to produce it, but chose to do so in PVC for reasons of cost and because the steel version was hard for users to handle. Kartell sold the Bookworm in a range of lengths and colours, and their mass-produced versions remained true to Arad's original idea – a bookcase that users could configure themselves to fit their own space requirements in whatever fluid, curving shape best suited their taste.

Arad was already well known in the design community for his individually made, hand-welded sheet-metal furniture and lighting that reflected his post-industrial aesthetic. The Bookworm marked his entry into design for mass-production, and its success means that the piece is still made today. Together with other furniture created for Kartell, Vitra, and similar companies, Arad's body of work and original way of thinking about design has a large and appreciative following.

### RON **ARAD**

#### 1951–

Born in Tel Aviv, Ron Arad attended the Jerusalem Academy of Art before moving to London to study at the Architectural Association. In 1981, with Caroline Thorman, he founded One Off in London's Covent Garden to design, produce, and sell handmade items, including the Rover chair. In the 1990s, he pursued collaborations with major manufacturers, such as Kartell and Alessi, as well as moving into the field of interiors, designing numerous shops as well as a foyer for the Tel Aviv Opera House. He was the architect for the Design Museum Holon, Israel, which opened in 2010.

## ON **DESIGN**

Ron Arad first attracted attention at the beginning of the 1980s for recycling "found" materials in creative ways. His most celebrated early product was the Rover chair, made from old leather seats from Rover cars set on curvaceous tubular frames. Aerial - a light mounted on a car radio antenna – appeared at around the same time. The industrial materials and aesthetic of these objects made them look at home with the high-tech designs fashionable at the time, but their reuse of old components was very different from high tech's love of beautifully finished and specially fabricated components of stainless steel and glass.

▲ **Rover chair**, Ron Arad, 1981

# Visual tour

**KEY**

▶ **BRACKET** The brackets attached to the Bookworm at regular intervals have two purposes. They act as bookends, keeping the books in place, and they incorporate fixings that enable the user to hang the shelf on the wall. Their position on the wall dictates the Bookworm's shape.

▶ **UNDERSIDE** The Kartell Bookworm is made of translucent PVC, so the uprights (and any books placed on the shelf) are visible through the material. Together with the deep colour, this translucency gives the Bookworm a visual richness that is different from any other bookshelf or wall decoration, and that also contrasts strongly with the original sprung-steel design that Arad produced.

1

2

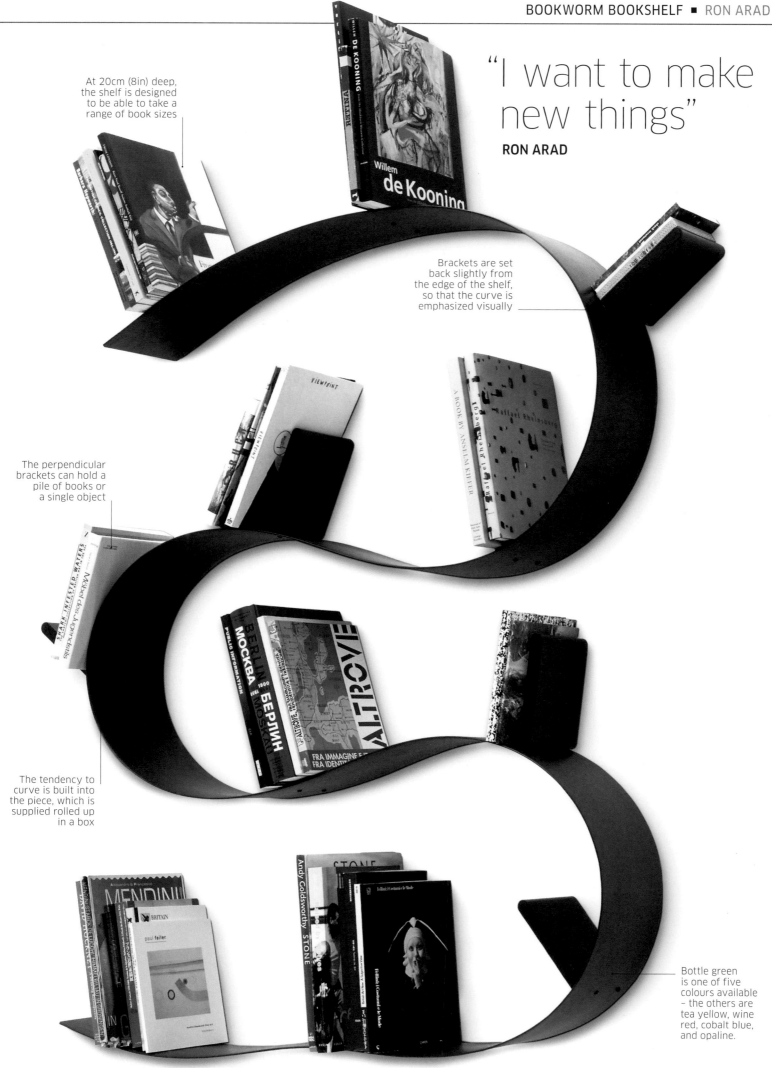

At 20cm (8in) deep, the shelf is designed to be able to take a range of book sizes

## "I want to make new things"

**RON ARAD**

Brackets are set back slightly from the edge of the shelf, so that the curve is emphasized visually

The perpendicular brackets can hold a pile of books or a single object

The tendency to curve is built into the piece, which is supplied rolled up in a box

Bottle green is one of five colours available – the others are tea yellow, wine red, cobalt blue, and opaline.

# 85 Lamps chandelier

1993 ■ LIGHTING ■ VARIOUS MATERIALS ■ THE NETHERLANDS

SCALE

## RODY GRAUMANS, DROOG DESIGN

**The 85 Lamps chandelier**, created by Dutch designer Rody Graumans, reduces the multiple light fitting to its bare bones. In its original version it consisted simply of 85 incandescent light bulbs, their plastic lampholders, and 85 wires joined in a cluster at the top. By combining these everyday objects in this inventive way, Graumans' chandelier forces people to look at them anew, so that they can appreciate the formal beauty and purposeful design of the light bulb, for example. It also embodies the way in which many Dutch designers were working in the 1990s: examining ideas or products that seemed old-fashioned or worn out and giving them a new lease of life; exploring concepts of recycling; and bringing a thought-provoking wit to their products.

The chandelier first appeared in an exhibition at the Milan Furniture Fair in 1993 that included a range of items by young Dutch designers. Like the other exhibits, the chandelier embodied the very qualities of simplicity and dry humour that inspired the founders to call this group of designers "Droog" (the Dutch word for "dry"). Their work was well received, but one criticism levelled at the chandelier was that its 85 lamps (although of low wattage) used a lot of electricity. This charge has been answered in the most recent version, in which the manufacturers have replaced the incandescent bulbs with economical LED lamps – a change that is true to Droog's original focus on both recycling old ideas and making things new.

### DROOG **DESIGN**

#### 1993–

The concept of Droog Design was born in 1993, when product designer Gijs Bakker and design historian Renny Ramakers collaborated on an exhibition at the Milan Furniture Fair. The following year they created the Droog Design Foundation and secured a deal with a manufacturing and distribution company. Further exhibitions followed, involving a number of independent designers who were producing work that reflected the Droog ideals of originality, clear concepts, and a humorous, no-nonsense approach. In 2003 the group set up its own production and distribution company, and shortly afterwards launched a shop and gallery in Amsterdam. Droog designers continue to initiate projects, branching out into areas such as shop interiors and kitchen design, and have taken commissions from high-profile clients including Levi Strauss and British Airways. They also mount exhibitions, organize seminars, and publish widely, helping to keep their innovative ideas and lateral thinking at the forefront of the design community.

> "Less and more are united in a single product"
> **DROOG DESIGN**

### ON **DESIGN**

Much of the work of Droog Design is low-tech, and involves bringing together inexpensive, standardized, or recycled materials in inventive ways. As well as the 85 Lamps chandelier, early Droog items included Tejo Remy's chests, made by bundling together several old wooden drawers, and his rag chairs, which consist of discarded clothes and fabric remnants, layered and bound with strips of steel. Applying upmarket design and production skills to the lowliest of materials, the Dutch group has taken similar ideas to the developing world. For example, vases by Nadine Sterk and Lonnie van Rijswijck encase recycled plastic containers in handcrafted wooden forms made by artisans in São Paulo, who benefit from the work and income.

▶ **Rag chair,** Tejo Remy, 1991

The original chandelier used 15-watt light bulbs; these have now been replaced with 1.5-watt LED lamps

Plastic wire connectors, usually hidden from view, become a visual feature, breaking up the tangle of wires at the top of the chandelier

The wires converge towards the ceiling, creating a shape that tapers overall

# Visual tour

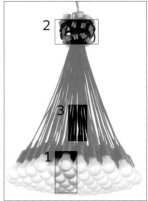

**KEY**

▶ **LAMPS** Seeing the bare bulbs without any covering or shade enables the viewer to appreciate their clean design at close quarters. Viewed from below, the mass of rounded, glowing forms functions as a single light as well as a dramatic focal point.

1

2

▲ **CONVERGENCE POINT** In contrast to the clarity of the lower part of the chandelier, the top of the piece is an apparently tangled confusion of wires and plastic connectors. This cluster of wires and connections forms a roughly spherical knot at the top of the chandelier, acting visually like a finial.

▶ **WIRES** In most multi-bulb light units the wires are concealed inside the fitting, and a single cable links the fitting to the ceiling. In the 85 Lamps chandelier, by contrast, each bulb has its own wire. This exposed wiring is an example of the way the Droog designers like their objects to be clear cut and easy to understand.

3

# Vermelha chair

1993 ■ FURNITURE ■ ALUMINIUM, STEEL, AND COTTON ■ BRAZIL

SCALE

## FERNANDO AND HUMBERTO CAMPANA

**When the first Vermelha (red) chair** appeared in 1993, it caused a stir. With its hundreds of loops of thick red rope, it was unlike any other chair before, and when the prototype was displayed as part of a gallery show in São Paulo, Brazil, many people considered it to be an individual and rather bizarre work of art rather than a piece of furniture. The designers – brothers Fernando and Humberto Campana – themselves emphasized the chair's symbolic meaning, describing it as a portrait of the chaotic melting pot that is their home country.

Manufacturers were unsure how to actually make the Vermelha chair, but eventually the Italian company Edra took it on. The chair's structural components are a metal frame consisting of three legs, a flat plate forming the seat, and a series of uprights. The manufacturer builds up a basket-like web of rope between the uprights to form the back and then overlays this with a series of carefully woven loops of thread, which are also built up over the seat. The arrangement of the upholstery looks random, but the loops are in fact carefully organized and the painstaking weaving process takes around 50 hours for each chair. The overlapping layers of rope make the chair both robust and comfortable, but it is perhaps the striking, slightly surreal, appearance of the piece which has made it the Campana brothers' bestselling product, and a symbol of the inventiveness of Brazilian design.

## Visual tour

**KEY**

➤ **ROPES** The upholstery is built up of handwoven loops formed from a continuous length of around 500m (1,640ft) of rope. This is made from an acrylic rope core wrapped around with red cotton thread. These two materials give the upholstery both its strength and softness.

1

2

◄ **BACK** The chair back is supported by nine aluminium uprights that are joined to the plate that forms the seat. The uprights protrude beyond the top of the woven ropes: their brushed finish adds a touch of brightness to the chair, and they also provide a sense of the underlying structure of the piece.

---

### FERNANDO AND HUMBERTO **CAMPANA**

**1961- AND 1953-**

When Fernando Campana qualified as an architect in the early 1980s, he opened a studio with his elder brother, Humberto, who had studied law. The two began to develop a series of chairs, mainly based on waste and recycled materials. Initially, their work was not widely known outside Brazil, but the publicity surrounding the Vermelha chair and a 1998 exhibition at New York's Museum of Modern Art raised their profile considerably. The brothers are now celebrated for their uncompromising product designs and sculptures.

### ON **DESIGN**

Cardboard, scraps of fabric, and rubber hose are among the low-cost or recycled materials that the Campana brothers use in their furniture. Their work defies categorization, and although their pieces may seem arbitrarily arranged or rough-edged, they are meticulously researched and crafted. The Favela chair (1991), created from wood offcuts, makes reference to Brazil's urban shanty towns, while the Banquete (2002) is built up from children's soft toys. The Sushi chair (2002), made of rags and remnants, looks like a basket of patchwork pieces.

➤ **Sushi chair**, Fernando and Humberto Campana, 2002

Overlapping loops of rope form dense upholstery

▲ **Side view**

# "...materials tell us the extent to which they can and want to be transformed"

**HUMBERTO CAMPANA**

Loops of rope are allowed to hang down to emphasize the random appearance of the weaving

The legs slant inwards to meet the concealed seat plate

▲ **Front view**

# Dyson DC01 vacuum cleaner

1993 ■ PRODUCT DESIGN ■ PLASTICS AND METAL ■ UK

SCALE

## JAMES DYSON

**In 1978, British engineer and inventor James Dyson** noticed that his vacuum cleaner was losing suction. When he opened it up, he saw that the machine separated the dust from the air by drawing air in through the sides of the porous bag, leaving the dust inside. The bag soon clogged up however, which resulted in poor suction. At around the same time, Dyson visited a sawmill, where he saw a different system of expelling waste. The mill used a cyclone, which spun air around a large cone to remove the sawdust by centrifugal force. Dyson reasoned that if he added a cyclone mechanism to a vacuum cleaner, there would be no need for a bag at all and the suction problem would be solved.

In 1985, after years of work, a revolutionary, bagless cleaner designed by Dyson went on the market in Japan. This model, called the G-Force, provided the funds for Dyson to develop his idea further, and in 1993 he launched the DC01 in the UK. This was an improved version of the original cyclone cleaner and people were not only impressed with its efficiency but also by its startling appearance. With its grey and yellow, partly transparent plastic body, Dyson's machine looked totally unlike the other vacuum cleaners available at the time, and its striking high-tech design was not only practical, but a huge marketing asset. The Dyson DC01, launched at the end of an economic recession, soon became the UK's bestselling vacuum cleaner and seemed to symbolize the country's regeneration through industrial innovation and design.

Integral carrying handle

▲ **Side view**

This mechanism allows the upper part of the body to tilt backwards

The cylinder is made of tough ABS plastic

The dust collects in this transparent cylinder

The suction is effective right up to the side, for cleaning around edges

## JAMES **DYSON**

### 1947–

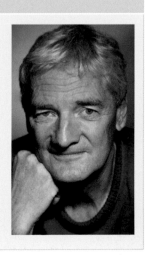

After studying furniture and interior design at the Royal College of Art, London, James Dyson began his career designing interiors and furniture for theatres, shops, and London's Heathrow Airport. In the 1970s, he set up his own company to develop products such as the Ballbarrow, an innovative wheelbarrow with a spherical wheel. Since producing his range of vacuum cleaners, Dyson has created a wide variety of other products that successfully combine advanced technology with cutting-edge design. They include washing machines, fans, the Airblade hand dryer, and the Airblade tap/dryer.

# Visual tour

> **CYLINDER** In spite of market research advice that buyers would be put off by a transparent cylinder in which they could see the dust, Dyson insisted on this feature. Seeing the dust spinning around, users appreciate how the cyclone works and can tell at a glance when the cylinder needs to be emptied.

**KEY**

> **TOOL STORAGE** On-board tool storage was not a common feature of vacuum cleaners when the DC01 was launched. The tools are stowed on the machine itself so that they are easy to find. Items such as this brush are in the same grey plastic as the body of the machine, and their lines follow similar patterns to those on the body.

< **COLOUR** By making the moving parts of the vacuum cleaner bright yellow, Dyson combined practicality with a strong sense of design. The grey and yellow gave the vacuum cleaner a unique character that appealed to customers and spawned many imitators.

## ON **DESIGN**

James Dyson tends to develop his products over long periods of time, and starts with pencil drawings rather than planning his designs on a computer. Sheets of sketches are then developed into models and prototypes. Creating the bagless vacuum cleaner involved more than 5,000 prototypes, none of which performed quite as the designer wanted. Dyson sees such difficulties in a positive light; "All failures are valuable because they all teach you something. I have lots of them every day." Although he faced many of his early failures alone, Dyson now has a huge research and development team working on new designs.

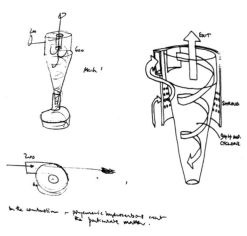

▲ **Sketchbook** showing the development of Root Cyclone™ technology, James Dyson, 1980s

## "I just want things to work properly"

**JAMES DYSON**

## IN **CONTEXT**

The striking combination of a transparent cylinder and a body made of grey plastic, with the key elements picked out in yellow, are signature Dyson features. Many later Dyson models – the low-slung cylinder designs, and subsequent uprights – have used a similar colour scheme. The more recent range of hand-held cleaners, however, represents a new departure, both in terms of technology and design. The DC44 Digital Slim is a cordless vacuum cleaner with a perfectly balanced, slender, electric blue aluminium wand. It weighs a mere 2.3kg (5lbs). By redesigning the vacuum cleaner Dyson have once again changed the way to clean, making it easy to vacuum both the ceiling and the floor, as well as the areas in between.

▲ **DC02**, 1995    ▲ **DC08**, 2002    ▲ **DC25**, 2008    ▲ **DC44**, 2012

# Verdana typeface

1994 ▪ GRAPHICS ▪ UK/USA

## MATTHEW CARTER

**Until the late 20th century, typography** was handled by the printing industry, which produced paper-based products such as newspapers and books. The rise of personal computers in the 1980s and 1990s, however, led to a radical transformation of the whole industry. Not all traditional typefaces display well on a computer screen and there was a demand for new fonts specially designed with screens in mind. In 1994, Virginia Howlett of Microsoft commissioned a font from Matthew Carter, a British typographer. He worked with type editor Thomas Rickner and two years later the font was released as Verdana. Carter designed the typeface to work well in small sizes and with low-resolution printers as well as ON screens. Its wide letter spacing makes it easy to read and the font also works well in many different languages. Bundled with computer operating systems, Verdana is now one of the world's most widely used fonts, its clarity and wide availability much prized by website and software designers. Sometimes it is also used for conventional graphics. In 2009, the Swedish retail giant Ikea decided to make the change from Futura to Verdana (which was already used on their website) in their printed catalogue. The choice of a font designed primarily for the screen proved controversial, but it is a clear demonstration of how, with internet use part of many people's everyday lives, Verdana has become ubiquitous.

ABCDEFGHIJKL
MNOPQRSTUVW
XYZabcdefghijk
lmnopqrstuvwx
yz0123456789

## MATTHEW **CARTER**

### 1937–

Born in London, Matthew Carter served an internship in a type foundry in the Netherlands. While there, he learned how to cut the steel punches that were used to make the matrices (moulds) to cast traditional metal type. He returned to London to work as a typographer, creating many typefaces for Linotype and designing the logo for the magazine *Private Eye*. In 1981, he co-founded the digital type foundry Bitstream Inc, and ten years later, he started another type foundry, Carter & Cone. He has designed numerous fonts for computer use, as well as type for publications including the *Washington Post*, *Boston Globe*, *New York Times*, and *Wired*.

## ON **DESIGN**

Fonts designed specially for computer equipment go back several decades. Arial was designed for office printers produced by IBM in 1982. It was the work of by a team at Monotype led by Robin Nicholas and Patricia Saunders. It is similar to Helvetica (see p.158), but has gentler curves and many of the letter terminations are diagonal rather than vertical or horizontal. Georgia, a serif font similar to Times New Roman, was designed for on-screen use by Matthew Carter in 1993. Compared to Times New Roman, Georgia has a greater x-height, looser spacing, slightly wider letters, and wider, blunter serifs. Comic Sans, designed in 1994 by Vincent Connare, imitates comic-book script and was intended as an informal font.

abcdefghijklmnopqrs
tuvwxyz1234567890

▲ **Georgia,** Mathew Carter, 1993

abcdefghijklmnopqr
stuvwxyz1234567890

▲ **Arial,** Robin Nicholas/Patricia Saunders, 1982

abcdefghijklmnopqrs
tuvwxyz1234567890

▲ **Comic Sans,** Vincent Connare, 1994

# Visual tour

**KEY**

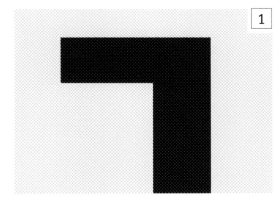

◀ **CAPITAL J** Care has been taken to ensure there is no confusion between similar letter forms. In Verdana, the top of the capital J has a small horizontal stroke, helping to differentiate it from the capital I and the number 1.

▶ **LOWER-CASE B** Matthew Carter designed Verdana with letters that have a generous x-height (the height of the lower-case x or curved part of the letter b). The letters are also set quite wide apart. This makes it less likely that they will run together when displayed on screen.

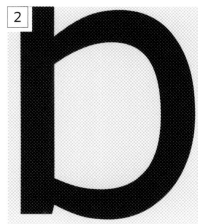

▲ **LOWER-CASE Q** The lower-case letter q is designed very simply, with a plain, vertical descender. However, closer inspection reveals the artful combination of straight lines and curves, together with the subtle variations in stroke width. This attention to detail shows Carter's typographic skill.

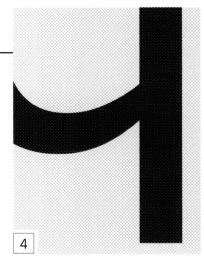

◀ **LOWER-CASE G** Most of the lower-case letters are relatively wide. This feature is clear from the swelling bowl of the lower-case g, which is broader than the tail immediately below it.

## 1996–present

- **Brick shelf system** Ronan and Erwan Bouroullec

- **Fjord Relax armchair** Patricia Urquiola

- **Garland light** Tord Boontje

- **Lover sofa** Pascal Mourgue

- **Miura stackable stool** Konstantin Grcic

- **Evolute table lamp** Matali Crasset

- **Spun chair** Thomas Heatherwick

- **Apple iPad** Jonathan Ive, Apple Industrial Design

- **Masters chair** Philippe Starck

# Brick shelf system

2001 ▪ FURNITURE ▪ POLYSTYRENE OR MDF ▪ FRANCE

SCALE

## RONAN AND ERWAN BOUROULLEC

**In the year 2000, the French designers** Ronan and Erwan Bouroullec were commissioned to work on the design of an exhibition of shoes for the Festival International des Arts de la Mode at Hyères in the South of France. Looking for a suitable display unit, they came up with the idea of producing a number of standardized elements that could easily be assembled and adapted to different exhibition sites. This was the origin of their ingenious Brick shelf system, which consists of modular components, like bricks, which can be stacked on top of each other and clipped together to form an open "wall" of shelving. By adding more units, the wall can be enlarged.

The Bouroullecs chose polystyrene as the material for the Brick shelving for two reasons – it has the smooth surface they required for the design, and it can be cut using a laser. The whole design follows a digital process, which the brothers have likened to a three-dimensional print-out: "This turns the laser principle into something like a big printer. The file leaves our studio, the manufacturer performs the cut-out process and then delivers the pieces". The system anticipated the 3D printing processes that have now become increasingly common.

Having designed an adaptable exhibition display stand and worked out how to produce it in a novel way, the brothers realized that they had created a flexible home-shelving unit that could also act as a stylish screen. The Italian company Cappellini adopted Brick for the domestic market and it was very well received. Its flowing, organic curves, practical design, and flexibility proved a great success with those who wanted an elegant structure on which to display decorative objects. Its clean lines and uncluttered appearance also meant that the Brick system was ideal as a partition that could be raised or lowered to divide space in modern, open-plan interiors.

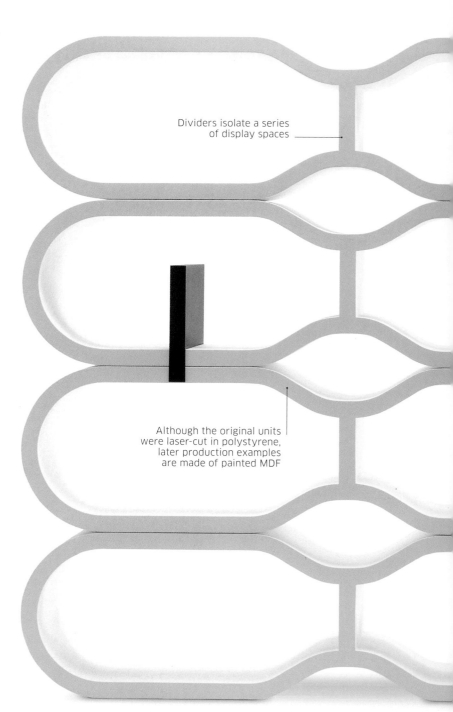

Dividers isolate a series of display spaces

Although the original units were laser-cut in polystyrene, later production examples are made of painted MDF

### RONAN AND ERWAN **BOUROULLEC**

#### 1971– AND 1976–

The Bouroullec brothers were born near Quimper, France. Ronan studied industrial and furniture design in Paris; Erwan attended the École Nationale Supérieure d'Arts de Paris, Cergy. Their first joint projects included a jewellery collection for SMAK Design, tableware for Habitat, and lights and ceramics for Cappellini. The brothers continue to design together and are much in demand. Their client list includes Vitra, Ligne Roset, Alessi, and Flos, and their work has been exhibited worldwide.

# Visual tour

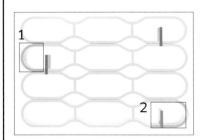

## KEY

**1**

**2**

▶ **SHELF ENDS** One of the distinctive features of the Brick system, as opposed to conventional shelving, is the semicircular shelf ends. Polystyrene and MDF were ideal materials with which to create the continuous, flowing lines across the base of the unit, around the sides, and along the top.

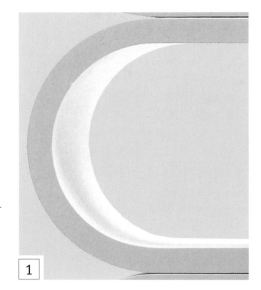

**1**

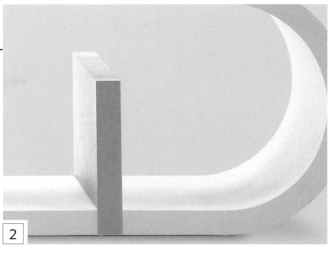

**2**

▲ **UPRIGHTS** When planning Brick as a display backdrop for an exhibition, the Bouroullecs wanted a clean, neutral surface that would not distract the eye from the exhibits. The white surface of the polystyrene is ideal, but a little relief is added by the coloured uprights, which act like visual punctuation marks, emphasizing the pure white of the rest of the unit.

Coloured uprights divide the compartment or support items placed on the shelf

Stacking one unit on top of another creates small spaces in which objects can be displayed

## ON **DESIGN**

Much of the Bouroullecs' work consists of items that can be assembled in different ways. This tendency began early on with a set of Vases Combinatoires that Ronan Bouroullec designed in 1997. These polypropylene vases, which were made in sets of eight, can be joined to one another to form vessels of different shapes and sizes. The brothers elaborated on this idea on a much larger scale with the Brick shelf system, and with other projects, such as the Cloud shelf system, which also consists of small modules that can be stacked and aligned. Even more ambitious is the Polystyrene House, a series of polystyrene strips and wooden sections that can be slotted together to make a house. Designs like these demonstrate the brothers' fascination with the architectural aspect of product design and their desire to make users think about how they use and interact with the objects around them.

▲ **Cloud shelf system**, Ronan Bouroullec, 2004

# Fjord Relax armchair

2002 ▪ FURNITURE ▪ STEEL, POLYURETHANE FOAM, AND FABRIC ▪ ITALY

## PATRICIA URQUIOLA

SCALE

**When Spanish-born designer Patricia Urquiola** went to Scandinavia, she was impressed by the modern furniture she saw there. A design that particularly caught her eye was the famous Egg chair, by Arne Jacobsen (see p.162). Urquiola decided to design an armchair inspired by the Egg, which would take its curvaceous, enveloping form in a new direction and reflect the quality of the dramatic scenery she saw in Scandinavia.

Her armchair has rounded contours like the Egg, but it features a deeply cleft back reminiscent of Norway's fjords, which give Urquiola's chair – and the accompanying chairs, stools, and footstools – their name. The underlying lightweight steel structure, covered with polyurethane foam and finished with fabric or leather covers, has a dramatic, lopsided form with an organic quality. If the chair is shaped like an egg, it is a broken egg. Urquiola was conscious of this and wanted to express the beauty of incomplete objects, such as broken shells, or pebbles on a beach that have been eroded by the waves. She also bore in mind that people today sit in a more relaxed and informal manner than they did in the late 1950s, when Jacobsen designed the Egg. In Urquiola's chair, you can sprawl and hook one leg over the arm, yet still be comfortable. To create the Fjord, Urquiola worked, as she often does, with her colleague Patrizia Moroso, whose furniture company's craftsmanship ensures that her distinctive chairs are manufactured and finished to the highest standard.

### PATRICIA **URQUIOLA**

**1961–**

Born in Oviedo, Spain, Patricia Urquiola studied architecture in Madrid, then trained in Milan. Her thesis was supervised by Achille Castiglioni (see p.180), who fuelled her interest in product design. Urquiola became an assistant lecturer on some of Castiglioni's courses, then worked on shops, restaurants, and other projects with a firm of architects. In 1996, she became head of the Milan design group, Lissoni Associati, and also worked independently on product design. Her designs include ceramics, textiles, and lighting, as well as her celebrated furniture.

# Visual tour

**KEY**

▶ **CLEFT** This split in the back of the chair is the Fjord chair's distinguishing feature. Although the split reduces the back area of the chair, the swelling curves in the lower part of the back, which extend to form the arms, provide good lumbar support.

1

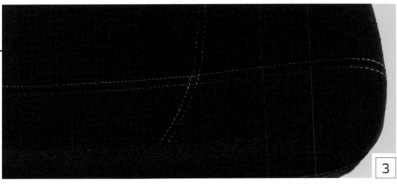

3

▲ **UPHOLSTERY** The visible stitching adds a contemporary touch to the seat cover. Although covers in a plain colour are popular, the Fjord is also made with covers in contrasting colours for the outer and inner parts of the chair.

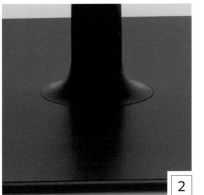

2

◀ **BASE** The chair sits on a strong metal pedestal. The single spindle seems to support the chair as if it were floating in the air, adding to its unconventional appearance.

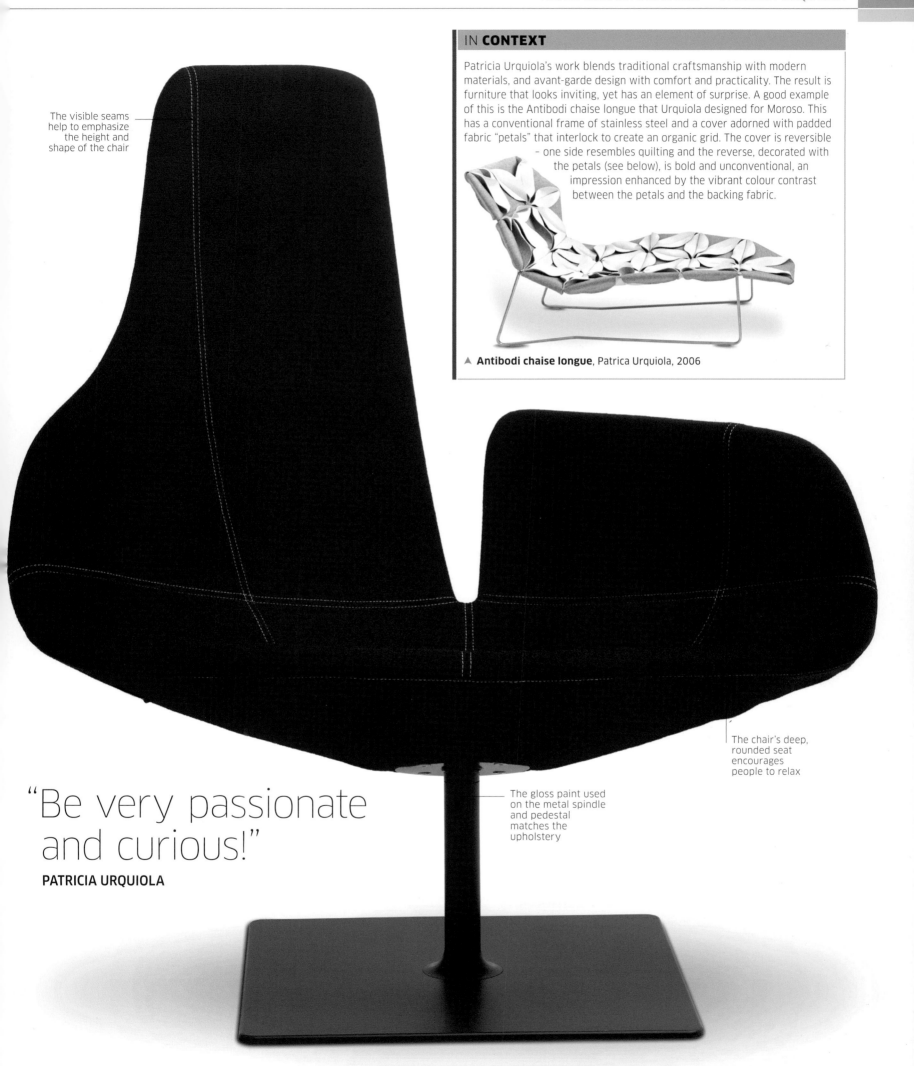

The visible seams help to emphasize the height and shape of the chair

Patricia Urquiola's work blends traditional craftsmanship with modern materials, and avant-garde design with comfort and practicality. The result is furniture that looks inviting, yet has an element of surprise. A good example of this is the Antibodi chaise longue that Urquiola designed for Moroso. This has a conventional frame of stainless steel and a cover adorned with padded fabric "petals" that interlock to create an organic grid. The cover is reversible – one side resembles quilting and the reverse, decorated with the petals (see below), is bold and unconventional, an impression enhanced by the vibrant colour contrast between the petals and the backing fabric.

▲ **Antibodi chaise longue**, Patrica Urquiola, 2006

The chair's deep, rounded seat encourages people to relax

The gloss paint used on the metal spindle and pedestal matches the upholstery

"Be very passionate and curious!"

**PATRICIA URQUIOLA**

# Garland light

2002 ▪ LIGHTING ▪ NICKEL-PLATED BRASS ▪ THE NETHERLANDS

## TORD BOONTJE

**Much of the work of Dutch designer** Tord Boontje shows that a contemporary approach and the latest technology do not have to result in cold, minimal design, but can produce objects that are rich and sensuous. One of the best examples of this is his Wednesday light, a garland of flowers and leaves made of sheet metal that is arranged around a lightbulb. Boontje designed it soon after the birth of his daughter, an event that encouraged him to explore more decorative and "feminine" themes.

The original Wednesday light was made of a sheet of stainless steel that was photo-etched into its intricate pattern of blossoms and foliage and wrapped around a lampholder and bulb. When the light was turned on, the shade glittered, sending a pattern of dappled light and shade across the room – the perfect response to the designer's love of the varying light within forests and the brilliant sparkle of ice and crystals. When the light came out in 2001, it attracted a lot of attention, and Boontje decided to produce a cheaper, mass-produced version – the Garland light – in softer nickel-plated brass that was sold in a flat-pack. The idea is that users can bend and shape the brass shade as they wish, so that each light becomes unique – they can even link several shades together to make a large, chandelier-like light fitting. This version of the light is a perfect example of the way that a designer's vision and modern manufacturing techniques can bring great design to people on a limited budget.

SCALE

The main stem has a series of tight curves that act as the light's structural skeleton

The lower elements also pick up light and reflect it around the room

The bottom of the fitting ends in a delicate trail of small flowers and leaves

## TORD **BOONTJE**

### 1968–

Tord Boontje studied industrial design at the Design Academy Eindhoven and did a Masters degree at the Royal College of Art, London. Having been based in both London and France, he returned to London in 2009 to take up the position of Professor of Design Products at the Royal College of Art. He has produced work – notably furniture, fabrics, lighting, and glassware – for a wide range of high-profile clients, including Alexander McQueen, Moroso, and Swarovski. His designs are held in numerous museums and collections and have won many awards.

# Visual tour

**KEY**

◄ **FLOWER** One of the main repeating motifs in the pattern is this flower. It appears in several different sizes and has petals that splay towards the outer edge. The larger examples have slender spikes between the petals. At first glance, the flowers seem to have six petals, but a closer look reveals that they have just five – the place where the sixth petal would be is occupied by the stalk that joins the flower to the main branch.

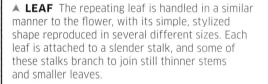

▲ **LEAF** The repeating leaf is handled in a similar manner to the flower, with its simple, stylized shape reproduced in several different sizes. Each leaf is attached to a slender stalk, and some of these stalks branch to join still thinner stems and smaller leaves.

▲ **LIGHT PATTERNS** The Garland light is made of one material – nickel-plated brass – in a uniform finish. However, the way that light from the bulb catches it means that some parts of the metal are brilliantly lit, while others are in dark shadow, producing a magical play of light across the whole piece.

▲ **LIGHT SOURCE** The metal flowers and leaves do not completely mask the light fitting. From some angles, it is possible to glimpse both the bulb and the lampholder, but these elements are mostly lost amongst the scintillating sparkle as your eye moves across the array of flowers and leaves.

## ON **DESIGN**

Tord Boontje started out at the interface of craft and design, making objects from recycled materials, such as a range of glass vessels that he produced from old wine and beer bottles. He has continued to draw inspiration from contemporary craft, and his work retains its simplicity and empathy with materials. When he created the Wednesday light, he said, "I didn't want to be too perfect or too fashionable... It's normal – like a Wednesday".

Works like the Wednesday and Garland lights take advantage of the precise photo-etching technology that has developed over recent decades. This method of working forges a direct line between design and

manufacturing in a way that appeals to designers such as Boontje, for whom the process of making has always held a strong fascination. Since the creation of the Wednesday light, Boontje has increasingly used new technology.

His interest in materials is evident in pieces such as the Blossom chandelier, which was made in response to a commission from Swarovski to rework the chandelier in a modern form. Boontje's reinterpretation is made up of clusters of crystal pieces arranged like tree blossoms and situated next to low-wattage LED lights. It forms a perfect marriage of old and new.

► **Blossom chandelier**, Tord Boontje, 2002

# Lover sofa

2003 ■ FURNITURE ■ MEMORY FOAM AND STAINLESS STEEL ■ FRANCE

SCALE

## PASCAL MOURGUE

**French designer Pascal Mourgue** conceived his Lover sofa and the other items in the Lover range of upholstered furniture, by looking at how the human hand can curl into a fist then uncurl again to open out flat. The furniture that he created based on this idea has the organic adaptability of the hand and provides the level of comfort that might be expected from a design based on human anatomy.

This innovative approach was possible because of a high-tech material called memory foam, a high-density foam – now commonly used in mattresses – which was originally invented at NASA's Ames Research Center for use in aircraft seats. Memory foam changes shape in reaction to pressure or the heat of the human body and, although some types return to their original shape when the pressure is released, extra-long-memory foam remains in place. Mourgue used this second type of foam for his sofas and chairs, which meant that the user could reshape parts of the seat. The sofa has a split back, so that each side of it can either be rolled up to make a cushion or positioned upright. Mourgue subsequently designed a matching bed, in which the foam rises up to form a high headboard. The sofa and the rest of the Lover range, all produced for Ligne Roset, are excellent examples of the blending of a modern, hi-tech material with innovative design.

◄ **Chaise longue** One end of the Lover chaise longue can be rolled over to make a supportive cushion that moulds itself to the user's head.

The wool and acrylic cover fabric was specially produced for this range

### PASCAL **MOURGUE**

**1943–**

Paris-born Pascal Mourgue studied sculpture at the Boulle school, then attended the École Nationale Supérieure des Arts Décoratifs there. In the early 1960s, while continuing to work as a fine artist, he began to design furniture, gradually turning his attention from office furniture to pieces for the home. He has had a long association with furniture design company Ligne Roset, for whom he designed the Lover, Calin, Sala, and other ranges. His diverse portfolio includes designs for ceramics, lighting, and sailing boats, as well as many paintings and sculptures.

# Visual tour

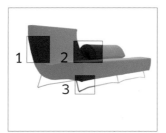

**KEY**

**► UPRIGHT BACK** The foam body narrows at the point where the base of the sofa curves upwards, to form the back. In its upright position, this back is as much about sheltering the sitter and defining personal space as it is about support.

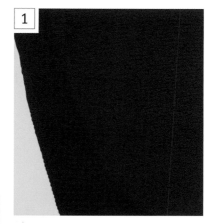

1

2

**▲ CUSHION** By rolling the back forwards, it becomes a supportive cushion that moulds itself to the body, because of memory foam's special characteristics. The cushion's rounded shape is supremely comfortable. It is also reminiscent of the baroque scrollwork that adorns 18th- and 19th-century sofas and chaise longues.

3

**▲ FOOT** The supports at the front and back of the sofa are made of brushed stainless steel attached invisibly to the frame. They do not detract from the focus of the design – the richly coloured seat and back.

## ON **DESIGN**

Pascal Mourgue's training as a sculptor is clear from many of his furniture designs. The seats, backs, cushions, and supports of his chairs and sofas have a sculptural quality, demonstrating both a creative handling of space and an open attitude to how people might choose to sit and relax. The Downtown range has movable backrests that can be positioned in different ways, sometimes at right angles to one another, so that people can change how they are sitting, depending on what they are doing or how they feel. Matching lumbar cushions can be used for support if required. The rounded shapes and soft edges contribute to the furniture's sculptural qualities, making the pieces both tactile and visually appealing.

**▲ Downtown sofa,** Pascal Mourgue, 2006

The high-density memory foam enables the cushion to stay rolled

The deep base and short legs mean that the sofa is set close to the floor

The sofa was made in red, and in six other colours

## "When it rolls up, you can relax in its embrace"

**PASCAL MOURGUE**

# Miura stackable stool

2003 ■ FURNITURE ■ FIBREGLASS-REINFORCED POLYPROPYLENE ■ GERMANY

SCALE

## KONSTANTIN GRCIC

**When Italian design company Plank** commissioned the German designer Konstantin Grcic to design a plastic bar stool, he used the opportunity to move away from the hard angles that had characterized his earlier designs. The stool had to be robust enough for commercial as well as domestic use, but its edges are soft and the design is more sculptural. The reinforced polypropylene makes the stool strong and suitable for use indoors and outdoors, but also very light so that it is easy to move around and stack.

Many bar stools are hard and unyielding, so Grcic paid special attention to comfort, sculpting the seat so that it rises slightly at the back, and tapers downwards at the front corners, rather like a saddle. The stretcher – which acts as both a footrest and a stacking device – also slopes down, echoing the line of the seat. The ergonomic design means the stool moulds comfortably to the contours of the body, and the polypropylene also flexes to accommodate a person's weight. The complex planes and soft, curving edges give the stool a fluid and perfectly balanced profile.

## "A design has to have a strong motivation for existing"

**KONSTANTIN GRCIC**, 1994

## Visual tour

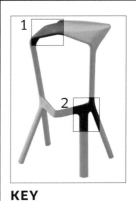

**KEY**

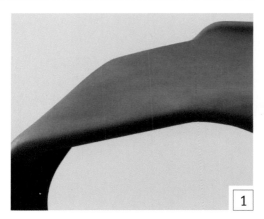

▲ **SEAT** The triangular sections at the front corners of the seat owe something to the planar geometry of Grcic's earlier furniture, but the edges are gently curved, making the stool both tactile and comfortable.

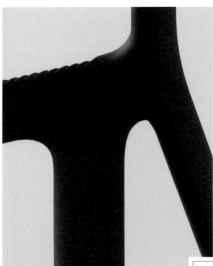

◄ **LEGS AND STRETCHER** About halfway up the stool there is a junction where the legs meet the horizontal stretcher and upright that supports the seat. This unusual arrangement works visually, partly because each piece meets its neighbours with a slight curve, giving the stool an organic, unified appearance.

## KONSTANTIN **GRCIC**

### 1965–

Born in Munich, Konstantin Grcic studied cabinet-making in the UK before taking a Master's in furniture design at London's Royal College of Art. He then spent a year in the studio of designer Jasper Morrison before setting up independently in Munich. His breakthrough came with the die-cast aluminium Chair One for Magis, which was followed by a wide range of furniture and product designs for companies including Cappellini, Flos, Lamy, and Muji. He also restyled the whole product range for Krups. Grcic's work has been exhibited widely and he has won many awards.

## IN **CONTEXT**

Konstantin Grcic has also designed a host of other simple yet multifunctional products. The Mayday lamp for Flos has a hanging hook and a plastic case which operates as both a diffuser (for ambient light) and reflector (for task lighting). Jet, for Agape, is a bathroom mirror with an ingenious swivel unit that reveals a storage unit. Likewise, the Coathangerbrush elegantly combines two related functions – hanger, and clothes brush.

▶ **Coathangerbrush**, Konstantin Grcic, 2004

The rear lip stops the user from sitting too far back

The slope on the front seat is echoed by a similar angle in the stretcher below

The strong material and balanced design make it possible to support the seat on slender uprights

The stretcher is angled towards the front to accommodate the front legs and balance the seat

The legs splay outwards to give stability

## ON DESIGN

For Grcic, who trained as a cabinet-maker, physically making things is very important. He likes to create an object by switching between his computer and 3D models. These may simply be made of cardboard at first, then sculpted in foam. Grcic paints the foam model, then finally produces a prototype in the intended materials. He has said that alternating between computer work and physical models helps him understand the complexity of the design and consider all the elements, from the materials to the user's needs, leading to solutions that might not be found using CAD (computer-aided design) alone.

▲ **Myto chair**, prototype and finished item, 2008

▲ **Stacking** The stools are easy to stack. This is especially useful for contract clients, who may need to store a large number of stools in a small space.

# Evolute table lamp

2004 ■ LIGHTING ■ CHROMIUM-PLATED STEEL AND MAPLE WOOD ■ FRANCE/ITALY

SCALE

## MATALI CRASSET

**French designer Matali Crasset** is known for her ability to look at objects with a fresh eye, coming up with innovative design concepts that make people think differently about what surrounds them. She also has a deep interest in transforming the environment, and in the way in which objects themselves can change form, especially from two to three dimensions. As she has said, "I love objects that, thanks to their folds, can transform and undergo an unexpected metamorphosis".

Crasset was an obvious choice for Italian company Danese, when they wanted to commission a new flat-pack light fitting. The surprise, however, was Crasset's counter-intuitive choice of wood as the material for her lamp shade. A length of thin maple veneer is cut precisely in such a way that, when folded and held in position with simple fasteners it forms a shape similar to a sphere. The wooden shade is then mounted on a minimalist stand of chromium-plated steel to make a beautifully proportioned lamp. When you switch on the lamp, the thin maple wood glows, displaying the material's natural grain and suffusing the space in rich, gentle light. The form of this deceptively simple lamp recalls paper origami models, yet the ingenuity of the concept and its delicate, sculptural quality are both testament to Crasset's thoughtful, artistic approach. The natural glow and softness of the wooden shade, each of which has a unique pattern, combines effortlessly and harmoniously with the hard shine of the metal stand. It is a supremely elegant and sensuous design.

### MATALI **CRASSET**

#### 1965–

After starting a course in marketing, Matali Crasset switched to design at the École Nationale Supérieure de Création Industrielle, Paris. After graduating, she trained with designer Denis Santachiara in Milan then returned to Paris to work with Philippe Starck and then for Thomson Multimedia. Since 1998, Crasset's own design studio has produced a range of lighting, furniture, and ceramics. Her domestic and commercial interiors include the temporary home of the Stedelijk Museum (SM's) in Amsterdam, for which she created the exterior, interior, website, and corporate identity.

# Visual tour

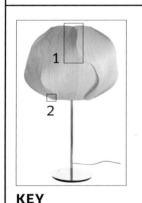

**KEY**

▶ **OVERLAPS** The maple shade has a series of paired holes. To transform the flat pack into a three-dimensional shade, each pair of holes is pulled together and fixed in place with a fastener. At these joining points the wood overlaps slightly, making the shade more opaque and varying the quality of the light.

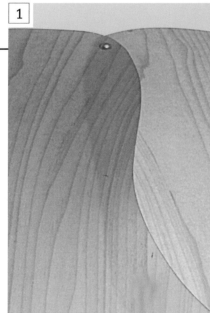

◀ **SUPPORT** Inside the wooden sphere a framework of metal struts, just visible at the base of the shade, holds the structure in place. Only slender metal components are required because of the relatively low weight of the thin maple. The whole lamp gives an impression of lightness and delicacy, like a piece of origami.

Variations in the wood grain make each lamp individual

"Thought always comes before image"

**Matali Crasset**

A slender steel tube supports the shade and bulb

The shiny base reflects some of the light from above, increasing the effect of the lamp

## IN **CONTEXT**

One of Matali Crasset's most celebrated interiors is the HI Hotel in Nice, France. For Crasset, a hotel – a place where guests are interested in trying a new kind of environment for a short time – is an ideal place for creating experimental interiors. Her approach was not to create a designed space in which everything is fixed and has a single, rigidly defined purpose. Instead, the HI Hotel has nine specific room concepts, each based on a different kind of spatial organization, with each room installation intended to offer the guest a range of possibilities for relaxation, conversation, work, and play. One room, for example, reflects cyber culture with giant pixels and an animated light box, while another resembles an indoor, decked terrace with planting.

▲ **HI Hotel**, Nice, Matali Crasset, 2003

## ON **DESIGN**

Typical of Matali Crasset's innovative way of rethinking interior spaces is her series of designs called Update/3 spaces. These were developed for the German company Dornbracht, which manufactures high-quality kitchen and bathroom fittings. Crasset's imaginative concepts for bathroom spaces are designed to offer us a new experience of bathing. Energiser, a space bathed in warm, yellow light, explores the pleasures of light and its effect on our wellbeing. In Phytolab, plants create a living wall that can screen the bathroom in an open-plan interior with limited space. There is no mirror; the focus is on an enjoyment of bathing through the senses of smell and touch rather than sight.

▲ **Phytolab**, Matali Crasset, 2002

# Spun chair

2009 ▪ FURNITURE ▪ SPUN METAL OR PLASTIC ▪ UK

SCALE

## THOMAS HEATHERWICK

**Thomas Heatherwick** is a British designer who has become celebrated for his original designs, such as the Olympic Cauldron at the 2012 London Olympic Games. He often draws his inspiration from production processes, and a striking case in point is his Spun chair, which was first produced using a technique called metal spinning. Metal spinning works by pressing a rotating piece of sheet metal against a former. It is usually used to produce items such as cookware or other vessels, and Heatherwick came across it as the way timpani – the large drums often used in orchestras – were made. Heatherwick produced the prototype Spun chairs by spinning sheets of steel or copper and forming them into parts which, when welded together, created a circular, wasp-waisted form with a narrow base, a wide lower rim, and an upper section with a hollow, rounded interior.

Set upright, the chair looks like a sculptural urn, but when the chair is tilted to rest on the base and lower rim, the hollow section is positioned at an ideal angle to act as a seat. Once sitting in the chair, the user can rotate, rock from side to side, or sit still; the sides of the hollow interior act as the chair's back and arms. Heatherwick successfully exhibited the steel and copper prototypes in London and then collaborated with furniture manufacturer Magis to make a larger production run of the chairs in rotation-moulded plastic. In both forms, the chairs have gained wide publicity, confirming Heatherwick's ability to think of traditional objects in startlingly new ways.

The moulded plastic body forms a richly textured surface

The inner part of the chair forms the seat

## THOMAS **HEATHERWICK**

### 1970–

London-born Thomas Heatherwick studied design at Manchester Polytechnic and the Royal College of Art (RCA). Soon after leaving the RCA, he won an award for a temporary structure on the façade of the prominent Harvey Nichols department store. In 1994, he set up a mixed-discipline studio – now called Heatherwick Studio – and has since produced a variety of works in the fields of architecture, design, and sculpture. From a beach café at Littlehampton to studios at the Aberystwyth Arts Centre, from a new bus for London to a handbag for Longchamp made of a single spiralling length of zip, Heatherwick's designs combine head-turning originality with a rigorous eye for function and form.

# Visual tour

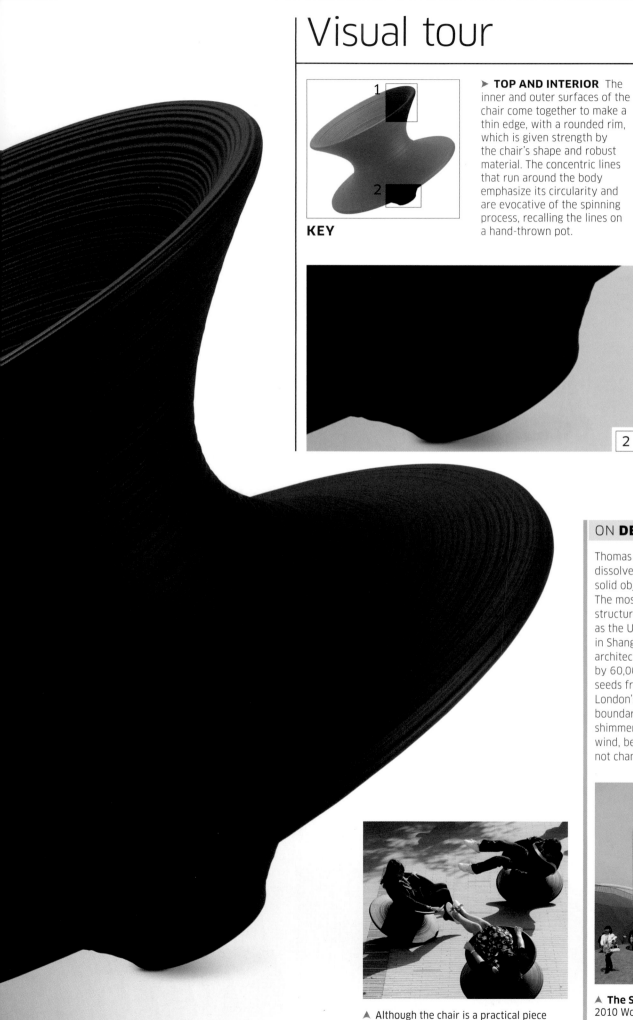

**KEY**

➤ **TOP AND INTERIOR** The inner and outer surfaces of the chair come together to make a thin edge, with a rounded rim, which is given strength by the chair's shape and robust material. The concentric lines that run around the body emphasize its circularity and are evocative of the spinning process, recalling the lines on a hand-thrown pot.

◄ **BASE** With its central rounded foot and swelling body, the chair looks rather like a child's old-fashioned spinning top. The base allows the chair to be tipped, as here, or left in an upright position.

## ON **DESIGN**

Thomas Heatherwick's work often seems to dissolve form and substance, making normally solid objects appear fragile and insubstantial. The most dramatic examples are his larger structures, such as the Seed Cathedral built as the UK Pavilion for the 2010 World Expo in Shanghai, which combines elements of both architecture and sculpture. Its walls are pierced by 60,000 clear acrylic rods, each containing seeds from the Millennium Seed Bank at London's Kew Gardens. The rods blur the boundary between the building and the sky, shimmer in the sun, and quiver gently in the wind, belying the notion that a building should not change or move.

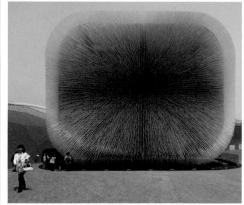

▲ **The Seed Cathedral (UK Pavilion),** 2010 World Expo, Shanghai, China

▲ Although the chair is a practical piece of furniture, it is also meant to be fun.

# Apple iPad

2010 ▪ PRODUCT DESIGN ▪ VARIOUS MATERIALS ▪ USA

SCALE

## JONATHAN IVE, APPLE INDUSTRIAL DESIGN

**In Spring 2010, Apple released the iPad**, the portable device that made the internet, email, photographs, music, and a variety of other functions available at the touch of a finger. Apple developed the iPad over many years, aiming to make a huge improvement on early tablet computers. These had been relatively heavy, ran a limited range of software, and mostly relied on the user activating the screen with a cumbersome, pen-like stylus. With its responsive, touch-sensitive screen, lighter weight, and array of apps, the iPad broke new ground, but what really set it apart and helped to make the product an instant success was its elegant, pared-down appearance. This was the work of Jonathan Ive, Apple's talented product designer, whose team had produced the iPod and iPhone, adopting a clean, functional aesthetic that recalls the work of German designer Dieter Rams (see p.144). The front of the iPad is uncluttered – there is only one button and most of the surface is taken up by the screen. The case tapers towards the edges, which emphasizes its slimness, and the smooth contours make the iPad very comfortable to hold. The simplicity of the design focuses attention on the display, with its ranks of icons. These icons, already familiar to many users from other Apple products such as the iPhone, are carefully designed to be instantly comprehensible to anyone who has not used an Apple device before. The icons, touch-screen, and sleek body come together perfectly in this elegant and user-friendly device that has changed the way people use and think about computers.

### JONATHAN **IVE**

**1967–**

British designer Jonathan Ive studied industrial design at Newcastle. He co-founded Tangerine, an industrial design agency, and was hired as a consultant by Apple, before working for them full-time. His first project for Apple was an update of the company's Newton MessagePad, an early hand-held computer. Under co-founder, Steve Jobs, Ive's team created the award-winning 1998 iMac, but in his subsequent work for the company (as Senior Vice president of Industrial Design), he has moved away from its exuberant colours, replacing them with the leaner lines and more sober colour palette of the iPhone and iPad.

## ON **DESIGN**

In 1998, Apple launched the first iMac, one of Jonathan Ive's most celebrated designs. Totally different from the box-like, visually neutral, white or beige computers of the time, the iMac was cased in brightly coloured translucent plastic. The machine not only looked user-friendly, it was designed to work straight out of the box, and connecting it to the internet only took two steps. The original iMac was an instant hit with consumers, and along with a number of equally brightly coloured successors, it sold in vast numbers. These sales transformed Apple – then in decline – into a successful, design-oriented global brand.

▲ **Apple "Bondi" iMac**, Jonathan Ive, 1998

# Visual tour

**KEY**

➤ **BUTTON** The single button on the front of the iPad bears a square screen icon. Pressing the button activates the display when the iPad is in sleep mode, and returns the display to the home screen when an app is running.

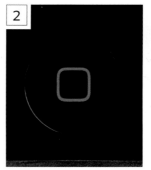

➤ **SCREEN AND FRAME**
The edge of the glass iPad screen forms a black border, which sets off the colour display without drawing attention to itself. The metal case that covers the back of the device curves up to protect the sides and edge of the screen.

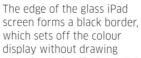

1

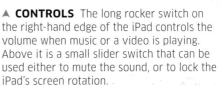

3

▲ **CONTROLS** The long rocker switch on the right-hand edge of the iPad controls the volume when music or a video is playing. Above it is a small slider switch that can be used either to mute the sound, or to lock the iPad's screen rotation.

# "It's very easy to be different, but very difficult to be better"

**JONATHAN IVE**

# Masters chair

2010 ▪ FURNITURE ▪ POLYPROPYLENE ▪ FRANCE

SCALE

## PHILIPPE STARCK

**When designing a familiar object**, such as a chair, many designers are tempted to ignore the history of design in order to try and come up with something entirely original. There is, however, an alternative approach – taken by the Postmodernists, for example (see p.214) – which involves not only being aware of design precedents, but also deliberately incorporating elements of them into the new design. The Masters chair, by the renowned French designer Philippe Starck, exemplifies this second approach.

Starck based his design on the contrasting shapes of three iconic modern chairs – the broad Tulip chair by Eero Saarinen, the taller Eiffel chair by Charles and Ray Eames, and the winged Series 7 chair by Arne Jacobsen. He juxtaposed the distinctive outlines of the chair backs and skilfully incorporated them into his own chair. These lines overlap and cross to form the sinuous back and arms of Starck's chair, before merging with the seat and legs.

The Masters chair weaves elements of the classic chairs into a coherent new design, with each one contributing a key element – the arms, the "wings", and the top of the chair back. This is a complex design, but the plain black polypropylene gives it visual unity, as does the uniform diameter of the three shapes. The result is a comfortable seat that is as light and practical as the 19th-century Thonet bentwood chair (see pp.14–15), to which Starck's chair also bears a passing resemblance. The chair was originally available only in black or white, but is now also produced in bright colours. With its numerous references to former icons, the Masters chair seems to encapsulate the history of modern chair design.

### PHILIPPE **STARCK**

**1949–**

French designer Philippe Starck set up his first studio in Paris at the age of 19. His career received a boost in 1982, when he designed the interior of President François Mitterand's private apartments. Interiors for high-profile hotels, cafés, and clubs soon followed. Since then, Starck has enjoyed a successful career as a designer of interiors and products. He has created many furniture designs, as well as a multitude of household objects, computer equipment, a motorcycle, and even a domestic wind turbine.

# Visual tour

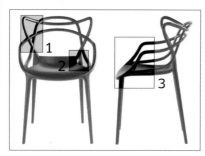

**KEY**

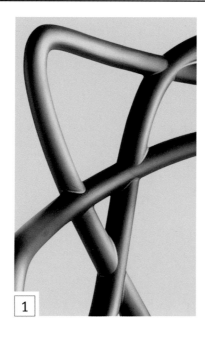

▶ **INTERSECTIONS** The way in which the lines of the three orginal chairs intersect creates an interesting pattern of geometric shapes that gives the Masters chair its unique character. The shapes are not just decorative, but help to give the chair strength and make it comfortable.

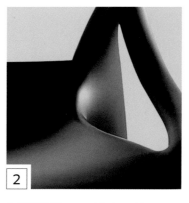

▲ **SEAT** The moulded seat is designed for optimum comfort. The edge of the seat is slightly raised at the back, helping to create a continuous, flowing line with the rear diagonals and legs. This feature also helps to provide additional lumbar support.

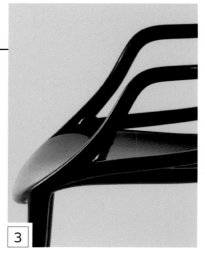

▲ **ARM** Starck exploited the sculptural potential of polypropylene to mould the gentle curves of the arms. Each chair arm is made up of two curves – one of them follows the outline of the Eames chair arm and the other is based on the arm of the Saarinen chair.

The "wing" echoes the outline of the Jacobsen chair

This section of the back follows the outline of the Eames chair

This part of the arm is based on Saarinen's chair design

The back leg forms a continuous line with the "wing"

▲ Side view

The legs broaden at the top, to provide additional strength.

## ON **DESIGN**

The clear, fluid forms of Starck's designs are often imaginative and quirky. The Juicy Salif lemon squeezer, with its spidery legs and streamlined, teardrop-shaped body, has a biomorphic quality. This is also apparent in Starck's bathroom taps for Axor, where the components branch and merge like parts of a tree. Starck has, however, also created rectilinear, geometric designs, such as his digital alarm clock for Telefunken, which is essentially a high-tech box, and his watches with rectangular faces for Fossil. The sheer variety of Starck's output has led some to accuse him of producing exclusive sculptures rather than designs, but he has clearly embraced the idea of democratic design, creating a range of fresh, affordable objects suitable for mass production.

"Historical lines overlie each other and intersect to finally yield an organic design closest to our body today"

**PHILIPPE STARCK**

▶ **Juicy Salif lemon squeezer**, Philippe Starck, 1990

# Index

## A

Aalto, Aino 76
Aalto, Alvar 76–77
Aarnio, Eero 177
Aberystwyth Arts Centre 244
AEG 32, 33
Aesthetic Movement 26
AGA cooker 42
Agape 240
Aicher, Otl 161, 204–05
Air France offices 56
aircraft 113, 175
Alessi 200, 214, 215, 217, 220, 232
Alexander McQueen 236
American Institute of Decorators 110
"Anti-Design" movement 192, 193
Apple 218
   iPad 246–47
Apple Industrial Design 246
Arabia ceramics company 118, 119
Arad, Ron 220–21
Architectural Association 220
*Arena* magazine 212
Ariel technique 143
Arntz, Gerd 205
Art Deco 38–41, 62, 63, 73, 74, 78, 79, 92
Art Directors' Club of New York
   exhibition (1934) 92
Art Nouveau 15, 24, 25, 26, 32, 40, 41, 52
Art Students League, New York 78
Artemide 200, 201
*Arts & Architecture* magazine 155
Arts and Crafts Movement 19, 30, 40, 46
Ashbee, C.R. 19
Associazione per Il Disegno Industriale 180
Athélia 92
Austin Seven Mini 168–71, 182
Auxiliary Territorial Service (ATS) 115
Avedon, Richard 92, 94
Axor 249

## B

Badovici, Jean 50, 51
Baekeland, Leo 74
Bakker, Gijs 272
Ballets Russes 92
Bang & Olufsen 198, 199
Barcelona International Exhibition 1929:
   German Pavilion 58
Barnack, Oskar 128
Barnsley, Sidney 19
Bass, Saul 136–37
bathroom mirror, Jet 240
bathroom spaces 243
Bauhaus 45, 50, 58, 59, 85, 102, 134,
   156, 204
   poster 46–49
Bayer, Herbert 49, 92–95

Beaton, Cecil 92
Beck, Harry 64–67
Behrens, Peter 32–33, 58, 59
Bel Geddes, Norman 90–91
Bellini, Mario 195
Benney, Gerald 123
Bernadotte, Sigvard 41
Bernadotte & Bjørn 198
Berthold 158
Bertoia, Harry 120–21, 150
Bertoni, Flaminio 138–41
Bialetti, Alfonso 62–63
Bialetti, Renato 62
Bianconi, Fulvio 106, 107
bicycles, Moulton 182–85
Bill, Max 156–57, 161
Bing and Grondal 41
Bitstream Inc 229
Bjerknes, Christian 68–69
BMW, Mini Hatch 168
BOAC Super VC10 aircraft interior 113
Bonet, Antonio 86–87
bookcase, Carlton 210–11
books, Penguin paperback covers 102–103
bookshelf, Bookworm 220–21
Boontje, Tord 236–37
*Boston Globe* 229
Botta, Mario 201
bottles
   L'Élégance perfume 38–39
   Telline perfume 39
Bouroullec, Ronan and Erwan 232–33
Brancusi, Constantin 104, 105
Brandt, Marianne 46
Braun 144, 145, 147, 166, 204
Breuer, Marcel 46, 50, 59, 75
Brionvega 188, 189, 200
British Motor Corporation (BMC) 168, 182
Brodovitch, Alexey 92–95
Brody, Neville 212–13
brooch 41
Brûlé, Tyler 161
brutalism 57
Buckminster Fuller, Richard 108
Buick 173–74
buses, Greyhound Scenicruiser 71
Byzantine art 187

## C

cabinets, Revolving cabinet 196–97
Cadillac, series 62 172–75
Cadillac, Antoine de la Mothe 173
calculators 195
cameras
   Kodak Bantam Special 78–79
   Kodak Instamatic 33 190–91
   Leica 129
   M3 Rangefinder 128–31

Campana, Fernando 224–25
Campana, Humberto 224–25
Campari 149
candelabra, Knuten 84–85
candleholder 41
Cappellini 196, 216, 232, 240
cars 191
   Austin Seven Mini 168–71, 182
   Cadillac Series 62 172–75
   Citroën DS 138–41
   Volkswagen Beetle Model 1300 80–83, 168
Carter & Cone 229
Carter, Matthew 228–29
Carwardine, George 72–73
Case Study Program 155
Cassandre, A.M. 54–55
Cassina 56
Castiglioni, Achille 180–81, 234
Castiglioni, Livio 180
Castiglioni, Pier Giacomo 180–81
chairs
   Ant chair 162
   Argyle chair 29
   Armchair No.31 77
   B3 (Wassily) chair 59
   Ball chair 177
   Banquete 224
   Barcelona chair 58–59
   B.K.F. Chair 86–87
   Chair One 240
   child's plastic stacking chair 188
   Cité armchair 52–53
   Cone chair 176, 177
   Diamond armchair 120–21
   Egg chair 9, 162–63, 234
   Eiffel chair 248, 249
   Embryo chair 217
   Favela 224
   Fjord Relax armchair 234–35
   folding campaign-style chair 87
   Hill House chair 28–29
   How High the Moon chair 219
   Lounge Chair 670 8, 132, 152–55
   Lover chair 238
   Masters chair 248–49
   Miss Blanche armchair 197
   Myto chair 241
   Panton chair 176–77
   Rag chair 224
   Red Blue chair 34–35
   rocking chair 155
   Rover chair 220
   S chair 176
   Series 7 chair 248, 249
   Snake chair 164
   Spun chair 244–45
   Sushi chair 224
   Swan chair 162
   Thonet Model 14 bentwood chair 14–15, 210
   Thonet Model 7500 rocking chaise 15

Tripolina chair 87
Tulip chair 150, 248, 249
Vermelha chair 224-25
Wiggle chair 202-3
Willow chair 29
Womb chair 150
Wood chair 216-17
chaises longues
    Antibodi 235
    LC4 56-57
    Lover 238
chandeliers
    85 Lamps 222-23
    Blossom 237
Chevrolet 173, 174
Chrysler 172, 175
Citroën 138-41
Citroën, André 138
Citterio, Antonio 195
classicism, Nordic 162
clocks
    Asterisk 109
    Atomic wall clock 108-9
    digital alarm clock 249
    Pod 217
    Sunburst 109
    wall clock 156-57
Coates, Wells 74-75
Coathangerbrush 240
cocktail shaker 41
coffee-maker, Moka Express 62-63
Compagnie des Wagon-Lits 55
Connare, Vincent 229
Conservatoire des Arts et Métiers, Paris 52
Constructivism 8, 44, 45, 161, 212
Copenhagen School of Applied Arts 198
Coty, François 38
Cranbrook Academy of Art, Michigan 120,
    121, 150, 152
Crasset, Matali 242-43
Cubism 51, 55, 148

**D**

Daimler Benz 188, 200
Dalén, Gustaf 42-43
Danese 242
D'Ascanio, Corradino 98-99, 100
Day, Lewis F. 17
Day, Lucienne 110-13
Day, Robin 110, 113
De Lucchi, Michele 195, 201
De Ponti, Luigi 62
De Prée, D.J. 108
de Stijl 34, 35
Deberny & Peignot type foundry, Paris 158
Deconstructivism 203
Dell, Christian 46
Design Academy Eindhoven 236
Design Centre 110
Design in Scandinavia exhibition 142
Design Laboratory 92, 94
Design Museum Holon, Israel 220
Deutscher Werkbund 32

dishes, Turned Leaf glass dish 117
Dobrolet airline 44
Dornbracht 243
D'Orsay 38
Dresser, Christopher 20-21
Driscoll, Clara 26-27
Droog Design Foundation 222-23
DuPont 78, 96
Dyson, James 226-27

**E**

Eames, Charles and Ray 109, 113, 120,
    121, 132, 150, 152-55, 248, 249
Eames House 155
Earl, Harvey 172-75
Eastman Kodak Company 79
École National Supérieure de Création
    Industrielle, Paris 242
École Nationale Supérieure d'Arts de Paris,
    Cergy 232
École Nationale Supérieure des Arts Décoratifs,
    Paris 238
Edra 224
E.K. Cole ("Ekco") 74
El Lissitsky 9, 45
Electrisk Bureau 68
Elsener, Karl 22-23
Embassy Court flats, Brighton 75
Emerson company 90, 91
Ericsson 68
Exner, Virgil 175
Exposition Internationale des Arts Décoratifs
    et Industriels Modernes (Paris, 1925) 38,
    39, 162, 165
    Pavillon de l'Esprit Nouveau 56

**F**

The Face magazine 212-13
Facetti, Germano 103
Fallingwater, Pennsylvania 86
Faulkner, Kate 19
Favrile glass 26, 27
Fender, Leo 124-27
Fender Stratocaster 124-27
Festival International des Arts de la Mode,
    Hyères (South of France) 232
Festival of Britain (1951) 9, 110, 113
    symbol 114-15
flatware
    Chinese Ivory 123
    Embassy range 122
    Josef Hoffmann and the Wiener Werkstätte
        30-31
    Konge (Acorn) 40
    Minimal 123
    Mono range 166, 167
    Pride cutlery 122-23
    Studio cutlery 123
    Thrift stainless-steel range 122
Flos 180, 217, 232, 240
FontWorks 212
Foster, Norman 183, 201

Franck, Kaj 118-19
Frank, Josef 84-85
Frutiger, Adrian 158
furniture
    Calin range 238
    chairs see chairs
    Co/Struc System 109
    Easy Edges series 202
    Ephesos range 195
    Experimental Edges 202
    Japanese 29
    oak sideboard 19
    office 109, 195
    Papp series 166
    Revolving cabinet 196-97
    Sala range 238
    tables see tables
Furniture in Irregular Forms series
    (Kuramata) 197
Futurism 148

**G**

Games, Abram 114-15
Ganz, Josef: May Bug 83
Garzanti 106
Gehry, Frank 202-3
Gehry House, Santa Monica 203
General Motors 90, 91, 173, 174
Gerrit Rietveld Academie, Amsterdam 34
Gestetner duplicating machine 70
Gill, Eric 102
Gimson, Ernest 19
Gismondi, Ernesto 201
"Glasgow Four, The" 28
Glasgow School of Art 28, 29
Glasgow tearooms 28, 29
glassware
    Finlandia range 207
    Tapio range 117
    Ultima Thule range 117
Globus department store, Zürich 158
Government School of Design, London 20
Grange, Kenneth 73, 190-91
Graumans, Rody 222-23
Graves, Michael 214-15
Gray, Eileen 50-51
Grcic, Konstantin 240-41
Great Exhibition (1851) 115
Greyhound Scenicruiser bus 71
Gris, Juan 55
Gropius, Walter 32, 46
Grupo Austral 86
Gugelot, Hans 144-47
Guggenheim Museum, Bilbao, Spain 202
Guild of Handicraft 19
Guimard, Hector 25
guitars, Fender Stratocaster 124-27

**H**

Haas 158, 166
Habitat 232
Hackman 116

Hald, Edward 143
hand dryer, Airblade 226
Hardoy, Jorge Ferrari 86-87
Harper, Irving 108
*Harper's Bazaar* magazine 92-95
Harvey Nichols department store 244
Heal's furnishers, London 110
Heatherwick, Thomas 244-45
Heatherwick Studio 244
Heiberg, Jean 68-69
Helsingborg 55 exhibition 142
Helsinki Central School of Industrial Arts 187
Helsinki College of Applied Arts 118
Henningsen, Poul 164-65, 176
Herbert Terry and Sons 73
Herman Miller company 104, 108, 109, 152, 155
Hessische Metallwerke 166
HI Hotel, Nice 243
hi-fi equipment, Beogram 4000 198-99
Hill House, near Glasgow 28, 29
Hitchcock, Alfred 134
Hoffmann, Eduard 158
Hoffmann, Josef 30-31
Honeyman and Keppie 28
Hornsea College of Art, London 212
Horta, Victor 25
House of Fiction travelling exhibition 216
Howard Miller Clock Company 108
Howlett, Virginia 228
Humber cars 168

I
Idée 217
Iittala glassworks 116, 117, 118, 206
Ikea 228
Inchbold, Peter 122
Innocenti, Fernando 99
Intercity 125 train 191
International Typographic Style 156
Isokon company 74, 75
Isola, Maija 186-87
Issigonis, Alec 158, 168-71
Ive, Jonathan 246-47

J
Jacob Jensen Design 198
Jacobsen, Arne 162-63, 176, 234, 248, 249
James Dixon and Sons, Sheffield 20
Japan Industrial Designers' Association 132
Japanese design 20, 28
Japanese Folk Crafts Museum, Tokyo 132
Japanese furniture 29
Jeanneret, Pierre 56-57
Jensen, Georg 40, 41
Jensen, Jacob 198-99
Jerusalem Academy of Art 220
Johnston, Edward 66, 67
Jugendstil 32
jugs, Euclid thermos jug 215

K
Kanazawa College of Art 218
Kandinsky, Wassily 46, 59
Karelia Region 187
Karhula and Iitala 76
Kartell 180, 188, 220
Kaufmann, Edgar, Jr. 86
*kawaii* (cute) products 219
Kawakubo, Rei 196
Kawasaki, Kazuo 218-19
Kazan School of Art, Odessa 45
Keler, Peter 49
Kelmscott Press 17
Kepes, Gyorgy 136
kettles
    Whistling Bird kettle 214-15
    electric kettle 32-33
King, Perry A. 192-95
Klaarhamer, P.J.C. 34
Klee, Paul 46
Klimt, Gustav 30, 31
Klotz, Mme B.J. 39
Knoll furniture company 104, 120, 150, 180
Knuten candelabra 84-85
Kodak
    Bantam Special 78-79
    Brownie 78
    Instamatic 33 190-91
Koppel, Eva and Nils 164
Krups 240
Kuramata, Shiro 196-97, 219
Kurchan, Juan 86-87

L
Labofa 198
Lalique, René 38-39
lamps
    1924 PH 165
    Akari 105
    Anglepoise 72-73, 200
    Arco 180-81
    Evolute table lamp 242-43
    Floor Lamp No.24 33, 85
    Gherpe 193
    Mayday 240
    PH Artichoke (Kogle) 164-65
    PH Contrast 165
    table lamp 49
    Tiffany lamp 26-27
    Tizio desk lamp 200-201
    Tolomeo 201
Lamy 240
Langelinie Pavilion, Copenhagen 164
laptop, Echos 20 195
Lawn Road flats, Hampstead, London 75
Le Corbusier 14, 32, 51, 52, 56-57, 74, 86, 108, 156, 162
League of Nations building, Geneva 56
Leica Camera AG 128-31
Leitz, Ernst 128, 130, 131
L'Élégance perfume bottle 38-39
lemon squeezer, Juicy Salif 249

lights 222
    Garland light 236-37
    Wednesday light 236
Linotype company 158, 229
Lissoni Associati 234
Littlehampton, West Sussex: beach café 244
Lockheed P-38 Lightning fighter aircraft 175
locomotives, S1 71
Loewy, Raymond 8, 70-71
logos
    Emerson 91
    Fender 127
    Iittala 206
    Lufthansa 161
    Moulton 183, 185
London College of Printing 65, 212
London Passenger Transport Board 66
London Press Exchange 115
London Transport 65
London Underground map 64-67
Longchamp 244
Lord, Leonard 169
Lubs, Dietrich 144
Lufthansa 204
    logo 161
Lundin, Ingeborg 142-43
Lunning Prize 142

M
Macdonald, Frances 28
Macdonald, Margaret 28, 29
Mackintosh, Charles Rennie 28-29
McNair, Herbert 28
magazines
    *Arena* 212
    *Arts & Architecture* 155
    *The Face* 212-13
    *Harper's Bazaar* 92-95
    *Neue Grafik* 156
    *Private Eye* 229
    *Ver Sacrum* (Sacred Spring) 31
    *Vogue* 94, 176
Magis 240, 244
Maison la Roche, Paris 56
Malinowski, Arno 41
Marber, Romek 103
Marimekko 187
Matisse, Henri 68
Mazza, Sergio 201
Mellor, Corin 123
Mellor, David 122-23
Memphis Group 192, 196, 197, 210-11
Metropolitan Museum Art School, New York 26
Meyer, Hannes 46
Microsoft 226
Miedinger, Max 158-61
Mies van der Rohe, Ludwig 32, 46, 52, 58-59
Milan fair (1946) 94
Milan Furniture Fair 220, 222
Milan Politecnico 180
Milan Triennale 110, 116, 142

Millennium Seed Bank, Kew Gardens, Surrey 245
Mingei movement 132
Mini (car) 168-71
Minimalism 196, 199
Miró, Joan 94, 113
Model, Lisette 94
Modernism 20, 34, 40, 41, 51, 52, 56-59, 65, 74, 75, 76, 85, 86, 92, 102, 104, 108, 110, 132, 147, 150, 158, 161, 162, 177, 190, 192, 193, 211, 214, 215
Moët et Chandon poster 24-25
Moholy-Nagy, László 46
Mondrian, Piet 34, 35, 155
Monotype 229
Moroso, Patrizia 234, 235, 236
Morris (car company) 168
Morris, Marshall, Faulkner & Co. (later Morris & Co.) 17
Morris, May 19
Morris, William 16-19, 30
Morrison, Jasper 240
Moscow metro map 67
Moser, Kolomon 30, 31
motor scooters
    Cushman 98
    Vespa 98-101, 183
Moulton, Alex 169, 182-85
Mourgue, Pascal 234-35
Mucha, Alphonse 24-25
Muji 240
Müller-Brockmann, Josef 161
Murano glassworks 106, 107
murrine technique 107
Museum of Chinese in America 191
Museum of Modern Art, New York 86, 104, 150, 152, 224

N

Nash, Arthur J. 27
National Bank of Denmark headquarters, Copenhagen 162
Nazi Party, Nazism 18, 46, 84, 102
Necchi 148, 149
Nelson, George 104, 108-09
Nemoy, Maury 136
Neue Grafik magazine 156
Neurath, Otto 65, 205
New South Wales Crafts Council 216
"New Typography" 46
New York Times 229
New York World's Fair (1939): General Motors pavilion 90, 91
Newson, Marc 216-17
Nicholas, Robin 229
Nike 191
Nielsen, Harald 41
Nizzoli, Marcello 148-49, 195
Noguchi, Isamu 104-05, 108, 109
Nord Express poster 54-55
Norwegian National Academy of Fine Arts 68
Nurse Bakelite Intercom 104
Nuutajärvi-Notsjö glassworks 118, 119

O

office cubicle 109
Olbrich, J.M. 31
Olivetti 148, 192-95
Olympic Games 132, 205, 244
    Munich pictograms (1972) 204-205
One Off 220
Op art 176
Orrefors glassworks 142, 143
ottoman (with Lounge Chair 670) 152

P

Paimio Sanatorium, Finland 76
Panton, Verner 176-77
paperweight, Antelope 39
Papst, Walter 147
Paris Métro 25, 65
Paris Motor Show 141
Paris World's Fair (1937): Finnish pavilion76, 77
Parma Accademia delle Belle Arti 148
pencil sharpener, Raymond Loewy 70-71
Penguin Books 191
    paperback covers 102-103
Penn, Irving 94
Pennsylvania Railroad: S1 locomotive 71
Pentagram 190, 191
perfume bottles 106
    L'Élégance 38-39
    Telline 39
Perriand, Charlotte 56-57
Philadelphia Museum School of Industrial Art 92
phonograph and transistor radio, portable (Model TP1) 145
Phonosuper SK4 radiogram 144-47
photo-etching 236, 237
Piaggio 98, 99, 100, 101
Piaggio, Enrico 98
Picasso, Pablo 55
Pick, Frank 65
pitchers 40-41
Plank 240
platter, birch 116-17
Polystyrene House 233
Ponti, Gio 106, 188, 200
Pop art 176, 177, 192, 211
Popova, Liubov 45
Porsche, 356 sports car 81
Porsche, Ferdinand 80, 81, 83
posters
    Bauhaus 46-49
    "Blonde Bombshell" 115
    Dobrolet poster 44-45
    The Man with the Golden Arm 136-37
    Milan Tourism 149
    Moët et Chandon 24-25
    Nord Express 54-55
    road safety 161
    Zodiac 25
Postmodernism 196, 197, 200, 211, 214, 215, 248
Pre-Raphaelites 17
Printex 187

Pritchard, Jack 75
Private Eye magazine 229
Propst, Robert 109
Prouvé, Jean 52-53, 56
Przyrembel, Hans 46
Public Services Building, Portland, Oregon 214

R

Raacke, Peter 166-67
Radical Design movement 193
radiograms, Phonosuper SK4 144-47
radios
    Beolit 400 199
    Brionvega TS 502 189
    Ekco radio AD65 74-75
    Emerson Patriot radio 90-91
    portable phonograph and transistor radio (Model TP1) 145
Ramakers, Renny 222
Rams, Dieter 144-47, 148, 246
Ratia, Armi 187
Rationalism 148
Ray, Man 95
Rem 180
Remy, Tejo 222
Renaissance 180
Research Studios 212
Rickner, Thomas 228
Rietveld, Gerrit 34-35, 176
Robsjohn-Gibbings, T.H. 104
Rodchenko, Alexandr 44-45, 212
Rohde, Gilbert 109
Rohde, Johan 40-41
Ronchamp chapel 57
Rosenthal 110, 116, 206
Rosselli, Alberto 200
Round Building, Derbyshire 123
Royal College of Art, Copenhagen 176
Royal College of Art, London 122, 212, 226, 236, 240, 244
Ruskin, John 17, 19

S

Saarinen, Eero 150-51, 152, 248, 249
Saarinen, Eliel 150
St Catherine's College, Oxford 162
St Louis Gateway Arch 150
Saint Laurent, Yves 55
Saks Fifth Avenue 191
Santachiara, Denis 242
Sapper, Richard 188-89, 200-01
Sarpaneva, Timo 206-07
SAS Royal Hotel, Copenhagen 162
Saunders, Patricia 229
Savoy restaurant, Helsinki 76
Scarpa, Carlo 106
Schlemmer, Oskar 48
Schmidt, Joost 46-49
Schmoller, Hans 103
School of Nancy 52
Schröder House, Utrecht 34, 35

Schröder-Schräder, Truss 35
sculptures, *Orpheus* 105
Seagram Building, New York 58
Secessionist Exhibition, 8th (Vienna) 28
Seed Cathedral (UK Pavilion), Shanghai World
    Expo (2010) 245
Seibel, Herbert 166
Serafini, Luigi 201
Sert, Josep Lluis 86
sewing machines, Mirella 148-49
Shanghai World Expo (2010): Seed Cathedral
    (UK Pavilion) 245
shelf systems
    Brick 232-33
    Cloud 233
sideboards, oak 19
Slade School, London 51
"Slav Epic" 24
SMAK Design 232
Snow, Carmel 92
Society of Industrial Designers 78
sofas
    Downtown 239
    Lover 238-39
Sottsass, Ettore 192-95, 201, 210-11
Starck, Philippe 242, 248-49
Stedelijk Museum, Amsterdam 242
Stepanova, Varvara 44-45, 212
Sterk, Nadien 222
Stölzl, Gunta 46
stools
    Butterfly stool 132-33
    Miura stackable stool 240-41
    stacking ("elephant") stool 132
Studebaker 70
Studio 2 system 144
Superstudio 193
Surrealism 55, 92, 94, 104
Svenskt Tenn 84, 85
Swarovski 236
Swedish Gas Accumulator Ltd 42
Swiss Army Knife 22-23
Swiss Cutlery Guild (later Victorinox) 22, 23
Swiss International Airlines 161
Swiss style 161

**T**

tables
    Coffee table 104-105
    E1027 occasional table 50-51
    Quaderna console 193
    Tulip table 150-51
tableware
    Kilta 118-19
    Suomi 206-207
Takefu Knife Village 218
Tangerine 246
tap/dryer, Airblade 226
tapestries 17, 19
Tatlin, Vladimir: *Monument to the Third
    International* ("Tatlin's Tower") 45
Tatra 80, 82, 83
Tea and Coffee Piazza set 215

Teague, Walter Dorwin 78-79
teapots 122
    silver 20-21
Tel Aviv Opera House, Israel 220
telephones
    Ericsson telephone DHB 1001 68-69, 188
    Grillo folding telephone 188-89
television 189
textiles
    Calyx furnishing fabric 110-13
    Flotilla 112
    Herb Antony 113
    Isosceles 113
    Unikko fabric 186-87
Thonet 14-15
Thonet, Michael 14-15
Thorman, Caroline 220
Tiffany, Louis Comfort 25, 26-27
Tiffany Glass Company 26, 191
toys, Rocking Sculpture 147
Tripolina chair 87
Tschichold, Jan 102
Tugendhat House, Brno, Czechoslovakia 58
Tupper, Earl 96-97
Tupperware 96-97
TWA Terminal building, New York 150
typefaces
    Akzidenz-Grotesk 156, 158, 160, 161
    Arial 229
    Bodoni Ultra Bold 102
    Braggadocio 195
    Comic Sans 229
    Didone 95
    Futura 228
    Georgia 229
    Gill Sans 102
    Helvetica 158-61, 229
    Horizontal 158
    Neue Haas Grotesk 158
    Pro Arte 158
    Sabon 102
    Times New Roman 229
    Universal 49
    Univers 158, 161
    Verdana 228-29
typewriters 148, 195
    Valentine typewriter 192-95

**U**

Ulm School of Design, Germany 144, 147,
    156, 166, 204
Unité d'Habitation, Marseille 56
Urquinaona Tower, Barcelona 86
Urquiola, Patricia 234-35
Utzon, Jørn 198

**V**

vacuum cleaners, Dyson 226-27
Van Gogh Museum, Amsterdam 34
van Rijswijck, Lonnie 222
vases
    Apple 142-43

Ariel No.534 143
Dolphin 39
Fazzoletto 106-7
Finlandia bark-pattern 207
Melon 142
Murrine 107
pear-shaped 41
Pezzato 107
Profiles 143
Savoy 76-77
Vases Combinatoires 233
Venice Biennale 1954: Dutch pavilion 34
Venini 116, 206
Venini, Paolo 106-107
*Ver Sacrum* (Sacred Spring) (journal) 31
Vespa (scooter) 98-101, 183
Vienna Ringtheater 24, 25
Vienna Secession 28, 30, 31
Viennese Academy 31
Viipuri Library (Vyborg, Russia) 76
Villa Savoye, Poissy, France 57
Viners of Sheffield 123
Vitra 176, 220, 232
Vitsoe 144
*Vogue* 94, 176
Volkswagen 80
    Beetle Model 1300 80-83, 168
Voysey, C.F.A. 19

**W**

Wagenfeld, Wilhelm 46, 49
Walker and Hall 122
wall coverings, fruit wallpaper 16-19
*Washington Post* 229
watches, Pod wristwatch 217
watering cans, enamelled metal 21
Webb, Philip 17
wheelbarrow, Ballbarrow 226-27
wheelchair, Carna 218-19
Whistling Bird kettle 214-15
Wiener Werkstätte (Vienna Workshops) 30-31
Winkreative 161
*Wired* magazine 229
Wirkkala, Tapio 116-17
Wolf, Herbert, 94
woodcuts 17, 19
world receiver, T1000 144
Worthington, Tom 110
Wright, Frank Lloyd 86

**Y**

Yanagi, Soetsu 132
Yanagi, Sori 132-33
Young, Edward 102-03

**Z**

Zanotta 180
Zanuso, Marco 188-89, 200
Zenger 23
Zenith 104
Zürich Kunstgewerbeschule 158

# Acknowledgments


**Dorling Kindersley would like to thank the following people for their assistance with this book:** Hannah Bowen, Georgina Palffy, and Satu Fox for editorial assistance; Diana Le Core for compiling the index; Jane Ewart and Steve Woosnam-Savage for design assistance; Gary Ombler and Sabrina Robert for additional photography; Liz Moore and Julia Harris-Voss for picture research; David Barton and Steve Crozier for digital image retouching.

**The publisher would also like to thank the following for their generosity in allowing DK access to their collections for photography:** Michael Woolf of Moulton Preservation; Sarah McCauley, Marketing Coordinator at Aram Store, London UK; Helen Palao, General Manager of Cassina S.p.A, London UK; Simone John, Press Officer at The Conran Shop, London UK; Jenny Hodge and Clare Lingfield at Leica Camera Ltd, London UK.

The publisher would like to thank the following for their kind permission to reproduce their photographs.

Key: a = above; b = below/bottom; c = centre; f = far; l = left; r = right; t = top

**1 Dorling Kindersley:** The Conran Shop: stockist and photography location. Tel: 0844 848 4000; www.conranshop.co.uk. **2–3 Dorling Kindersley:** Aram Store: Stockist and photography location (Wire Diamond chair by Harry Bertoia). **4 The Bridgeman Art Library:** Private Collection / Christie's Images (tr). **5 The Bridgeman Art Library:** Private Collection / The Estate of Raymond Loewy (ftl). **Dorling Kindersley:** Judith Miller / Lyon and Turnbull Ltd. (tr/Savoy vase). **Photo SCALA, Florence:** Museum of Modern Art (MoMA), New York (tl). **6 akg-images:** Mucha Trust (cr). **The Bridgeman Art Library:** Private Collection / The Stapleton Collection (tl). **Dorling Kindersley:** Archivio Storico Piaggio, Pontedera (bl); Judith Miller / David Rago Auctions / © ADAGP, Paris and DACS, London 2013 (bc). **Edra Spa:** (c). **Etsy, Inc.:** (br). **Marimekko:** (tr). **7 Dorling Kindersley:** Le Corbusier, Jeanneret, Perriand – LC4 – Cassina I Maestri Collection. Cassina is the only company in the world authorized to produce this chaise-longue since 1964, when the architect was still alive, working in close collaboration with the Le Corbusier Foundation in Paris and Pernette Perriand-Barsac. A farsighted choice which over the years has made it possible to explore and share the architects' knowledge thanks to a passionate and attentive philological reconstruction process./© FLC / ADAGP, Paris and DACS, London 2013 (tl). **© MOURON. CASSANDRE. Lic 2013-19-02-05 www.cassandre-france.com:** (bl). **8 Alamy Images:** Interfoto (tr). **The Bridgeman Art Library:** Private Collection / The Stapleton Collection (tl); Tiffany & Co./Private Collection (cr). **Decophobia.com:** G. Reboul (cl). **Dorling Kindersley:** Aram Store: Stockist and photography location (br); Judith Miller / Lyon and Turnbull Ltd (bl). **11 Edra Spa:** (bl). **Courtesy of Marc Newson Ltd:** (br). **Photo SCALA, Florence:** Metropolitan Museum of Art, New York / Art Resource (cr); Museum of Modern Art (MoMA), New York (cl). **V&A Images / Victoria and Albert Museum, London:** The Robin and Lucienne Day Foundation (tr). **12 Photo SCALA, Florence:** Museum of Modern Art (MoMA), New York. **13 The Bridgeman Art Library:** Private Collection / The Stapleton Collection. **14 akg-images:** (b). **14–15 Photo SCALA, Florence:** Museum of Modern Art (MoMA), New York (main image). **15 Photo SCALA, Florence:** Museum of Modern Art (MoMA), New York (bl). **Thonet GmbH:** (br). **16 The Bridgeman Art Library:** Private Collection / The Stapleton Collection. 17 **Getty Images:** Rischgitz/Hulton Archive (crb). **V&A Images / Victoria and Albert Museum, London:** (bc). **18–19 The Bridgeman Art Library:** Private CollectionThe Stapleton Collection (main image). **19 The Bridgeman Art Library:** Kelmscott Manor, Oxfordshire (cra); Private Collection (bc). **20–21 Michael Whiteway:** (main image). **21 Photo SCALA, Florence:** Museum of Modern Art (MoMA), New York (tc). **22–23 PHOTOPRESS / Victorinox:** (all images). **24 The Bridgeman Art Library:** Mucha Trust (bc). **24–25 akg-images:** Mucha Trust (main image). **25 akg-images:** Mucha Trust (br). **Corbis:** Gavin Hellier / Robert Harding World Imagery (bl). **26 Photo SCALA, Florence:** Metropolitan Museum of Art, New York (cra). **26–27 The Bridgeman Art Library:** Tiffany & Co. / Private Collection (main image). **27 Dorling Kindersley:** Judith Miller / James D. Julia Inc. (tl). **28 Alamy Images:** Pictorial Press Ltd (bc). **28–29 National Trust for Scotland:** (main image). **29 Alamy Images:** V&A Images (br). **Corbis:** Thomas A. Heinz (bl). **30 Getty Images:** Imagno / Hulton Archive (bl). **30–31 Photo SCALA, Florence:** Museum of Modern Art (MoMA), New York (main image). **31 akg-images:** Erich Lessing (bc). **32 akg-images:** Ullstein Bild (bl). **32–33 Alamy Images:** Interfoto (main image). **33 Electrolux:** (br / AEG logos). **34 Centraal Museum, Utrecht:** RietveldSchröderArchief / Pictoright / © DACS 2013 (clb). **34–35 Photo SCALA, Florence:** Museum of Modern Art (MoMA), New York / © DACS 2013 (main image). **35 Alamy Images:** Richard Bryant / Arcaid Images / © DACS 2013 (bc). **36** Dorling Kindersley: Aram Store: Stockist and photography location. **37 © MOURON. CASSANDRE. Lic 2013-19-02-05 www.cassandre-france.com. 38 Getty Images:** Albin Guillot / Roger-Viollet (clb). **38–39 Dorling Kindersley:** Judith Miller / David Rago Auctions / © ADAGP, Paris and DACS, London 2013 (main image). **39 Corbis:** Philip Spruyt / Stapleton Collection (br). **Dorling Kindersley:** Judith Miller / David Rago Auctions / © ADAGP, Paris and DACS, London 2013 (fbl, bl, bc). **40 The Royal Library, Copenhagen:** (bl). **40–41 Photo SCALA, Florence:** Metropolitan Museum of Art, New York/Art Resource (main image). **41 Dorling Kindersley:** Judith Miller / The Silver Fund (cra, bl, br, fbr); Judith Miller / Von Zezschwitz (fbl). **42–43 AGA Rangemaster Group PLC. 44–45 akg-images:** © Rodchenko & Stepanova Archive, DACS, RAO, 2013 (main image). **45 Alamy Images:** ITAR-TASS Photo Agency (bl). **The Bridgeman Art Library:** Private Collection (br). **46 Bauhaus-Archiv Berlin:** (br). **Getty Images:** Apic / Hulton Archive / © DACS (tr). **47 Photo SCALA, Florence:** Museum of Modern Art (MoMA), New York/© DACS. **48–49 Photo SCALA, Florence:** Museum of Modern Art (MoMA), New York /© DACS (main image). **49 Bauhaus-Archiv Berlin:** © DACS 2013 (tr). **Photo SCALA, Florence:** Museum of Modern Art (MoMA), New York (crb). **TECTA:** (bc). **50–51 Dorling Kindersley:** Aram Designs Ltd, holders of the worldwide licence for Eileen Gray designs. Stockist and photography location: Aram Store (main image). **51 Image supplied by Aram Designs Ltd, holders of the worldwide licence for Eileen Gray designs. Stockist and photography location: Aram Store:** (br). **This image is reproduced with the kind permission of the National Museum of Ireland:** Berenice Abbott (cla). **V&A Images / Victoria and Albert Museum, London:** This image is reproduced with the kind permission of the National Museum of Ireland (cra). **52 Photo SCALA, Florence:** BI, ADAGP, Paris (cr). **52–53 From the collection of the Vitra Design Museum:** © ADAGP, Paris and DACS, London 2012 (main image). **53 RMN:** Centre Pompidou, MNAM-CCI, Dist. RMN-Grand Palais / Jean-Claude Planchet / Georges Meguerditchian / © ADAGP, Paris and DACS, London 2012 (tr). **54–55 © MOURON. CASSANDRE. Lic 2013-19-02-05 www.cassandre-france.com:** (main image). **55 Getty Images:** Gaston Paris / Roger-Viollet (cra). **56 ADAGP: Banque d'Images, Paris 2012:** © ADAGP, Paris and DACS, London 2013 (bl). **56–57 Dorling Kindersley:** Le Corbusier, Jeanneret, Perriand – LC4 – Cassina I Maestri Collection. Cassina is the only company in the world authorized to produce this chaise-longue since 1964, when the architect was still alive, working in close collaboration with the Le Corbusier Foundation in Paris and Pernette Perriand-Barsac. A farsighted choice which over the years has made it possible to explore and share the architects' knowledge thanks to a passionate and attentive philological reconstruction process. / © FLC ADAGP, Paris and DACS, London 2013 (main image). **57 Alamy Images:** Schütze / Rodemann / Bildarchiv Monheim GmbH / © FLC / ADAGP, Paris and DACS, London 2013 (br). **58–59 Dorling Kindersley:** Aram Store: Stockist and photography location (main image). **58 Corbis:** Bettmann (clb). **59 Alamy Images:** G. Jackson/Arcaid Images (br). **61 V&A Images / Victoria and Albert Museum, London. 63 Bialetti Industrie S.p.A.:** (bc, br). **64–65** TfL from the London Transport Museum collection. The Underground map is reproduced from the Collections of the London Transport Museum © Transport for London. The Underground "roundel" logo is ® Transport for London (main image). **65 © 1965 Ken Garland:** (br). **66–67 TfL from the London Transport Museum collection:** The Underground map is reproduced from the Collections of the London Transport Museum, © Transport for London. The Underground "roundel" logo is ® Transport for London. (main image). **67 The Bridgeman Art Library:** Victoria & Albert Museum, London / The Stapleton Collection / © Transport for London (cra). **68 Gerson Lessa / Flickr:** (br/mouthpiece). **68–69 Photo SCALA, Florence:** DeAgostini Picture Library (main image). **69 Holger Ellgaard:** (tr). **70–71 The Bridgeman Art Library:** Private Collection / The Estate of Raymond Loewy. **70 Corbis:** Condé Nast Archive (bc). **71 akg-images:** (br). **72–73 The Bridgeman Art Library:** Private Collection / Christie's Images (main image). **73 Images courtesy of Anglepoise®:** (tr); (br). **74–75 V&A Images / Victoria and Albert Museum, London:** (main image). **75 Alamy Images:** Arcaid (br). **76 Alamy Images:** Pictorial Press Ltd (bl). **76–77 Dorling Kindersley:** Judith Miller / Lyon and Turnbull Ltd (main image). **77 Alvar Aalto Museum, Finland:** Iittala (www.iittala.com) / © DACS 2013 (bc). **Dorling Kindersley:** Judith Miller / Wiener Kunst Auktionen – Palais Kinsky (br). **78 Walter Dorwin Teague Papers, Special Collections Research Center, Syracuse University Library:** (bl). **78–79 RMN:** Centre Pompidou, MNAM-CCI, Dist. RMN-Grand Palais / Bertrand Prévost (main image). **79 Getty Images:** Science & Society Picture Library (br). **RMN:** Centre Pompidou, MNAM-CCI, Dist. RMN-Grand Palais / Bertrand Prévost (t/case closed and close-ups). **81 Getty Images:** Heinrich Hoffmann / Hulton Archive (cra). **83 Corbis:** Hulton-Deutsch Collection (br). **Technické muzeum Tatra:** (cra). **84 Getty Images:** Imagno / Hulton Archive (c). **84–85 Svenskt Tenn:** (main image). **85 Svenskt Tenn:** (tr). **86–87 Photo SCALA, Florence:** Museum of Modern Art (MoMA), New York (main image). **86 akg-images:** Album / Documenta (clb). **87 Getty Images:** Martin & Osa Johnson Archive (br). **88 Dorling Kindersley:** Aram Store: Stockist and photography location. **89 V&A Images / Victoria and Albert Museum, London:** The Robin and Lucienne Day Foundation. **90 Corbis:** Bettmann (bl). **90-91 Decophobia.com:** G. Reboul (main image). **91 Decophobia.com:** G. Reboul (tr). **Harry Ransom Center, The University of Texas at Austin:** The Edith Lutyens and Norman Bel Geddes Foundation Inc. (tl). **92 Getty Images:** Walter Sanders /Time & Life Pictures (br). **93-95 Paper Pursuits Fashion & Design Print Collectibles:** Harper's Bazaar (Hearst) / © DACS 2013. **94 Alamy Images**: Pictorial Press Ltd/ Harper's Bazaar (Hearst) / © The Richard Avedon Foundation (bc). **Paper Pursuits Fashion & Design Print Collectibles:** Harper's Bazaar (Hearst) / © The Richard Avedon Foundation (br). **95 Iconofgraphics.com:** Harper's Bazaar (Hearst) / © Man Ray Trust / ADAGP, Paris and DACS, London 2013 (br). **96–97 Photo SCALA, Florence:** Museum of Modern Art (MoMA), New York (bc/storage containers). **96 Press Association Images:** AP (bc). **Photo SCALA, Florence:** Museum of Modern Art (MoMA), New York (cr, tr). **97 Photo SCALA, Florence:** Museum of Modern Art (MoMA), New York (tl/2, ca, tl/shaker, clb/stacking cups, tr, tl/3, tl/1, tc).

**Science Museum / Science & Society Picture Library:** (br). **98-101 Dorling Kindersley:** Archivio Storico Piaggio, Pontedera (main image + details). **99 Archivio Piaggio "Antonella Bechi Piaggio":** (tr). **101** Getty Images: Popperfoto (br).**102 Alamy Images:** The Art Archive (bl). **102-103 PENGUIN and the Penguin logo are trademarks of Penguin Books Ltd.:** (main image). **103 Corbis:** Robert Estall / PENGUIN and the Penguin logo are trademarks of Penguin Books Ltd. (cra). **PENGUIN and the Penguin logo are trademarks of Penguin Books Ltd.:** (br/book covers). **104 Getty Images:** Eliot Elisofon / Time & Life Pictures/© The Isamu Noguchi Foundation and Garden Museum / ARS, New York and DACS, London 2013 (bc). **Herman Miller:** (tr/scale image). **104-105 Dorling Kindersley:** Aram Store: Stockist and photography location (main image). **105 Corbis:** Nathan Benn/Ottochrome / © The Isamu Noguchi Foundation and Garden Museum / ARS, New York and DACS, London 2013 (cra). **The Noguchi Museum:** © The Isamu Noguchi Foundation and Garden Museum / ARS, New York and DACS, London 2013 (br). **106 Archivio Fotografico VENINI S.p.A.:** (bc/portraits). **106-107 Collection of the Gemeentemuseum Den Haag:** (main image). **107 Getty Images:** A. Dagli Orti / De Agostini (bc, br). **108-109 The Bridgeman Art Library:** Private Collection / Christie's Images (main image). **108 Corbis:** Grigsby Cassidy / Condé Nast Archive (cl). **109 Dorling Kindersley:** Judith Miller / Wallis and Wallis (cra). Herman Miller: (br). **110 University of Brighton Design Archives:** (br). **111 V&A Images / Victoria and Albert Museum, London:** The Robin and Lucienne Day Foundation. **112-113 V&A Images/Victoria and Albert Museum, London:** The Robin and Lucienne Day Foundation (main image). **113 British Airways Speedbird Centre:** (br). **V&A Images / Victoria and Albert Museum, London:** The Robin and Lucienne Day Foundation (ca, cra). **114 V&A Images / Victoria and Albert Museum, London:** with the kind permission of the Estate of Abram Games. **115 Getty Images:** Hulton Archive (cr). **V&A Images / Victoria and Albert Museum, London:** with the kind permission of the Estate of Abram Games (b). **116-117 Photo SCALA, Florence:** Museum of Modern Art (MoMA), New York (main image). **116 Wirkkala archives:** photo Martti Ounamo (bl). **117 Dorling Kindersley:** Judith Miller / Bonhams, Bayswater (cra). **118-119 Design Museum, Finland:** Rauno Träskelin (main images). **118 Design Museum Finland:** Rauno Träskelin (1). **Fiskars:** Indav Oy / Timo Kauppila (bc). **119 Dorling Kindersley:** Judith Miller / Jeanette Hayhurst Fine Glass (tr). **120 Press Association Images:** AP /WFA/ © ARS, NY and DACS, London 2013 (bl). **120-121 Dorling Kindersley:** Aram Store: Stockist and photography location (main image). **121 Knoll Inc.:** © ARS, NY and DACS, London 2013 (br). **122-123 David Mellor Design:** (main image). **122 Guzelian photographers:** Steve Forrest (cra). **123 David Mellor Design:** (cra). **Dorling Kindersley:** Judith Miller / Graham Cooley (br). **124-125 The Bridgeman Art Library:** Private Collection / Christie's Images (main image). **124-127** FENDER®, STRATOCASTER® and the distinctive headstock designs commonly found on FENDER guitars are registered trademarks of Fender Musical Instruments Corporation, and used herein with express written permission. All rights reserved. FMIC is not affiliated in any way with author or publisher. **125 Getty Images:** Jon Sievert/Michael Ochs Archives (cra). **126-127 The Bridgeman Art Library:** Private Collection / Christie's Images (main image). **127 Photo courtesy of Fender Musical Instruments Corporation:** (br). Getty Images: Nigel Osbourne / Redferns (bl/Rhinestone). **129 Alamy Images:** Hire Image Picture Library (cra). **132-133 Photo SCALA, Florence:** Museum of Modern Art (MoMA), New York (main image). **132 Press Association Images:** AP / Kyodo News (bc). **134** akg-images: Sotheby's. **136-137 The Bridgeman Art Library:** Private Collection / Christie's Images / The Estate of Saul Bass (main image). **136 Photograph by Harrie Verstappen - www.thelooniverse.com:** (bl). **139 © Citroën Communication:** (cra). **141 Rex Features:** Sipa Press (br). **142-143** akg-images: Sotheby's (main image). **143 Dorling Kindersley:** Judith Miller / Bukowskis (bl). **Heritage Auctions:** (br). **144-145 Photo SCALA, Florence:** Museum of Modern Art (MoMA), New York (main image). **144 Courtesy of the Vitsœ archive:** Photo by Abisag Tüllmann (bc). **145 Photo SCALA, Florence:** Museum of Modern Art (MoMA), New York (tr). **146 Photo SCALA, Florence:** Museum of Modern Art (MoMA), New York (tr/cabinet detail, tl, details). **147 Used with the kind permission of Dieter Rams:** (cra). **Photo SCALA, Florence:** Museum of Modern Art (MoMA), New York (tl, bl). Wilkhahn: (br). **148-149 Photo SCALA, Florence:** Museum of Modern Art (MoMA), New York (main image). **148 Photo Ugo Mulas © Ugo Mulas Heirs. All rights reserved:** (clb). **149 Photo SCALA, Florence:** Museum of Modern Art (MoMA), New York (tl, needle mechanism, controls & nameplate). **150 Corbis:** Oscar White (bl). Dorling Kindersley: Judith Miller / Freeman's (br). **150-151 Dorling Kindersley:** Judith Miller / Lyon and Turnbull Ltd. (main image). **152 Corbis:** John Bryson / Condé Nast Archive (bl). **152-153 Photo SCALA, Florence:** Museum of Modern Art (MoMA), New York (main image, tc/scale picture & front view). **153 Herman Miller:** (tc). 156-157 Photo SCALA, Florence: Museum of Modern Art (MoMA), New York (details 4 & 5, key 6 & 7 + details 1, 2, 6 & 7). **154 Herman Miller:** (tl/key - back). Photo SCALA, Florence: Museum of Modern Art (MoMA), New York (key - front). **155 The Bridgeman Art Library:** Saint Louis Art Museum, Missouri, USA / Gift of Mrs. Charles Lorenz (br). **Getty Images:** Peter Stackpole / Time & Life Pictures (cra). **156** akg-images: Imagno (bl). **Photo SCALA, Florence:** Museum of Modern Art (MoMA), New York / © DACS 2013 (br). **156-157 Photo SCALA, Florence:** Museum of Modern Art (MoMA), New York (main image). **161 Deutsche Lufthansa AG:** (cra). **Getty Images:** Tim Boyle (ca). **Photo SCALA, Florence:** Museum of Modern Art (MoMA), New York/© DACS 2013 (br, fbr). **162 Getty Images:** Reg Birkett (bc). **162-163 Dorling Kindersley:** The Conran Shop: stockist and photography location. Tel: 0844 848 4000 - www.conranshop.co.uk (main image). **164-165 Photo SCALA, Florence:** Museum of Modern Art (MoMA), New York (main image). **164 Scanpix Denmark:** Per Pejstrup (br). **165 Vintage Gallery Denmark:** (br). **166 Roman Raacke:** (br). 166-167 mono - a brand of Seibel Designpartner: (main image). **167 Photo SCALA, Florence:** Museum of Modern Art (MoMA), New York (br). **168 Getty Images:** Derek Berwin (bl). **169 British Motor Industry Heritage Trust:** (cra). **171 BMW AG:** (crb, br, fbr). Wikipedia: Bull-Doser (fcrb). **173 Getty Images:** Joseph Scherschel / Time & Life Pictures (tr). **Used with permission, GM Media Archives:** (cr). **175 Alamy Images:** Scott Germain / Stocktrek Images, Inc. (br); Motoring Picture

Library (bl). **176-177 V&A Images / Victoria and Albert Museum, London:** (main image). **176 Verner Panton Design:** (bl). **177 Adelta International:** (tc). **178 Photo SCALA, Florence:** Museum of Modern Art (MoMA), New York. **180-181 Dorling Kindersley:** Aram Store: Stockist and photography location (main image). **180 © Foundation Achille Castiglioni:** (br). **181 © Foundation Achille Castiglioni:** (cra). **182 Getty Images:** Keystone /Hulton Archive (bc). **183 Moulton Bicycle Company:** (cra). **186-187 Marimekko:** (main image). **187 Design Museum, Finland:** (bc). **Marimekko:** (br). **188-189 Photo SCALA, Florence:** Museum of Modern Art (MoMA), New York (main image). **188 Courtesy of Richard Sapper:** (bl). **189 Alamy Images:** INTERFOTO (cra). **190 Press Association Images:** PA Archive (cra). **191 Getty Images:** Science & Society Picture Library (cla). **Pentagram Design:** (cra). **192 Getty Images:** Keystone-France (clb). **Photo SCALA, Florence:** Museum of Modern Art (MoMA), New York (tr). **192-193 Photo SCALA, Florence:** Museum of Modern Art (MoMA), New York (main image). **193 The Bridgeman Art Library:** Private Collection / Photo © Bonhams, London (tr). **Photo SCALA, Florence:** Museum of Modern Art (MoMA), New York (br). **194 Photo SCALA, Florence:** Museum of Modern Art (MoMA), New York (cla, tl/key - side views, tl/key - 3/4 angle view, br). **V&A Images / Victoria and Albert Museum, London:** (tr). **194-195 Etsy, Inc.:** (branding & keys). **VintageEuroDesign:** Anouschka. **195 Alamy Images:** Marc Tielemans (br). **Photo SCALA, Florence:** Museum of Modern Art (MoMA), New York (clb). **V&A Images / Victoria and Albert Museum, London:** (bl, tl). www.vintagecalculators. com: (cra). **196 akg-images:** CDA / Guillemot (bc). **196-197 Cappellini Design Spa:** (main image). **197 Photo SCALA, Florence:** Museum of Modern Art (MoMA), New York (bl, br). **198-199 Photo SCALA, Florence:** Museum of Modern Art (MoMA), New York (main image). **198 Jacob Jensen Brand ApS:** (br). **199 Bang & Olufsen UK Ltd:** (crb, tc/Beovox 2500 Cube speaker). **Photo SCALA, Florence:** Museum of Modern Art (MoMA), New York (cr, tr, br). **200-201 The Bridgeman Art Library:** The Israel Museum, Jerusalem, Israel (main image). **201 akg-images:** (bl). **202 Corbis:** Roger Ressmeyer / Fish & Furniture © Frank Gehry & New City Editions (br). **202-203 Photo SCALA, Florence:** Museum of Modern Art (MoMA), New York (main image). **203 Alamy Images:** Natalie Tepper / Arcaid Images (tl). **204 Press Association Images:** Sven Simon / DPA (bc). **204-205 © 1976 by ERCO GmbH:** (main image). **205 Foundation Gerd Arntz Estate:** © DACS 2013 (tr/isotypes). **Mexico 1968 Sports Pictogram designers:** Lance Wyman, Beatrice Colle & Manuel Villazon: (cla/Mexico 1968). **206 Sarpaneva Design Oy:** Markku Luhtala (bl). **206-207 © Rosenthal GmbH GERMANY:** (br/Suomi tableware). **207 Dorling Kindersley:** Judith Miller / The Glass Merchant (cra). **208 Courtesy of Marc Newson Ltd. 209 Photo SCALA, Florence:** Metropolitan Museum of Art, New York/Art Resource. **210-211 Photo SCALA, Florence:** Metropolitan Museum of Art, New York /Art Resource (main image). **211 eyevine:** Contrasto (cra). **212 Getty Images:** Andy Short / Computer Arts Magazine (cra). **212-213 Alamy Images:** Antiques & Collectables / Bauer Media / Photography Jamie Morgan, Styling Ray Petri for Buffalo. Model Lindsey Thurlow (main image). **214 Corbis:** Christopher Felver (br). **214-215 V&A Images/Victoria and Albert Museum, London. 215 Courtesy of Michael Graves Design Group:** (br). **216-217 Courtesy of Marc Newson Ltd:** (main image). **216 Courtesy of Marc Newson Ltd:** Brett Boardman (bl). **217 Courtesy of Marc Newson Ltd:** (br); Fabrice Gousset (tl). **218-219 Photo SCALA, Florence:** Museum of Modern Art (MoMA), New York / Kazuo Kawasaki Direction, Ouzak Design Formation (main image). **218 Kazuo Kawasaki Direction, Ouzak Design Formation:** (bl). **219 Photo SCALA, Florence:** Museum of Modern Art (MoMA), New York (tr). **220 Alamy Images:** Steve Speller (cr). **Ron Arad Associates:** (bl). **220-221 V&A Images / Victoria and Albert Museum, London:** (main image). **222-223 Droog Design:** (main image). **222 Droog Design:** (bl). **224-225 Edra Spa:** (main image). **224 Campana Design Ltda EPP:** Fernando Laszlo (bc). Edra Spa: (br). **226-227 Dyson:** (all images). **229 Getty Images:** S. Byun / The Boston Globe (cra). **230 © KGID. 231 Studio Tord Boontje. 232 Alamy Images:** Danny Nebraska (bc). **232-233 Cappellini Design Spa:** Ronan & Erwan Bouroullec (main image). **233 Ronan et Erwan Bouroullec:** (br). **234 Getty Images:** Tullio M. Puglia (cra). **234-235 Photo SCALA, Florence:** Museum of Modern Art (MoMA), New York (main image). **235 Moroso S.p.A.:** Alessandro Paderni (tr). **236-237 Studio Tord Boontje:** (main image). **236 Studio Tord Boontje:** Phil Sayer (bc). **237 Studio Tord Boontje:** (br). **238-239 Ligne Roset:** (all images). **240 Corbis:** Tibor Bozi (bc). **© KGID** (br). **240-241 Photo SCALA, Florence:** Museum of Modern Art (MoMA), New York (main image). **241 © KGID** (br). **Photo SCALA, Florence:** Museum of Modern Art (MoMA), New York (fcra). **242-243 Danese, Milano:** © ADAGP, Paris and DACS, London 2013. **242 Danese, Milano:** © ADAGP, Paris and DACS, London 2013 (bc). **Getty Images:** Bertrand Rindoff Petroff / French Select (bl). **243 Corbis:** Sergio Pitamitz / © ADAGP, Paris and DACS, London 2013 (tr). **Dornbracht:** Uwe Spoering/© ADAGP, Paris and DACS, London 2013 (br). **244-245 Photo SCALA, Florence:** Museum of Modern Art (MoMA), New York (main image). **244 Corbis:** Colin McPherson (bc). **245 Getty Images:** View Pictures / UIG (br). **Susan Smart Photography:** (bc). **246 Corbis:** Patrick Fraser (cr). **Getty Images:** Bloomberg (bl). **248-249 Kartell Spa:** (main image). **248 Alamy Images:** Ludovic Maisant / Hemis (cl). **249 Kartell Spa:** (tr). **Photo SCALA, Florence:** Museum of Modern Art (MoMA), New York (br).

All other images © Dorling Kindersley
For further information see: www.dkimages.com